Letts

Revise A2

OCR
including Salters
Chemistry

Contents

Chapter 4 Chemistry of organic functional groups

Chapter 5 Polymers, analysis and synthesis

Chapter 6 Synoptic assessment

Specification lists

OCR A2 Chemistry

MODULE	SPECIFICATION TOPIC	CHAPTER REFERENCE	STUDIED IN CLASS	REVISED	PRACTICE QUESTIONS
A2 Unit F324 (M4) *Rings, polymers and analysis*	*Arenes*	4.6, 4.7, 4.8			
	Carbonyl compounds	4.2			
	Carboxylic acids and esters	4.3, 4.4, 4.5			
	Amines	4.9			
	Amino acids and proteins	5.1, 5.2			
	Polyesters and polyamides	5.2			
	Synthesis	5.6			
	Chromatography	5.3			
	Spectroscopy	AS 4.7, 5.4, 5.5			
A2 Unit F325 (M5) *Equilibria, energy and elements*	*How fast?*	1.1, 1.2			
	How far?	1.3			
	Acids, bases and buffers	1.4, 1.5, 1.6			
	Lattice enthalpy	2.1			
	Enthalpy and entropy	2.2			
	Electrode potentials and fuel cells	2.3, 2.4			
	Transition elements	3.1, 3.2, 3.3, 3.4, 3.5, 3.6			

Examination analysis

A2 Chemistry comprises two unit tests. All questions are compulsory.
Synoptic assessment will form part of both units. Practical and investigative skills will also be assessed.

Units 1, 2 and 3 comprise AS Chemistry			50%
Unit F324	*Structured questions: short and extended answers*	*1 hr test*	15%
Unit F325	*Structured questions: short and extended answers*	*1 hr 45 min test*	25%
Unit F326	*Internal assessment of practical and investigative skills*		10%

Salters A2 Chemistry

MODULE	SPECIFICATION TOPIC	CHAPTER REFERENCE	STUDIED IN CLASS	REVISED	PRACTICE QUESTIONS
A2 Unit F334 **(M4)** *Chemistry of materials*	*Equilibria*	*1.4*			
	Organic functional groups	*4.1, 4.2, 4.3, 4.4, 4.5, 4.6, 4.7, 4.8, 4.9, 5.1, 5.2*			
	Organic reactions	*4.2, 4.3, 4.8, 5.1, 5.2*			
	Reaction mechanisms	*4.2, 4.9*			
	Applications of organic chemistry	*5.6*			
	Modern analytical techniques	*5.3, 5.4*			
	Application	*4.3, 4.4, 4.5, 4.9, 5.2*			
	Kinetics	*1.1, 1.2, 5.2*			
	Isomerism	*4.1*			
	Formulae, equations and amount of substance	*1.6, 3.4*			
	Redox	*2.3, 2.4, 3.2, 3.4*			
	Inorganic chemistry and the Periodic Table	*3.1, 3.3, 3.5, 3.6*			
A2 Unit F335 **(M5)** *Chemistry by design*	*Bonding and structure*	*4.9, 5.2, 5.6*			
	Rates of reaction	*1.1, 1.2*			
	Equilibria	*1.3, 1.4, 1.5, 1.6, 2.2, 2.4*			
	Redox	*2.3, 2.4*			
	Inorganic chemistry and the Periodic Table	*2.4*			
	Organic functional groups	*4.4, 4.6*			
	Organic reactions	*4.7, 4.9, 5.6*			
	Modern analytical techniques	*3.2, 4.9, 5.3, 5.4, 5.5*			
	Energetics	*2.1, 2.2*			
	Reaction mechanisms	*4.2*			
	Isomerism	*4.1*			
	Applications	*5.6*			

Examination analysis

A2 Chemistry comprises two unit tests. All questions are compulsory.
Synoptic assessment will form part of both units. Practical and investigative skills will also be assessed.

Units 1, 2 and 3 comprise AS Chemistry			*50%*
Unit F334	*Structured questions: short and extended answers*	*1 hr 30 min test*	*15%*
Unit F335	*Structured questions: short and extended answers*	*2 hr test*	*20%*
Unit F336	*Individual investigation*		*15%*

AS/A2 Level Chemistry courses

AS and A2

All Chemistry A Level courses currently studied are in two parts, with three separate modules in each part. Students first study the AS (Advanced Subsidiary) course. Some will then go on to study the second part of the A Level course, called A2. Advanced Subsidiary is assessed at the standard expected halfway through an A Level course: i.e., between GCSE and Advanced GCE. This means that new AS and A2 courses are designed so that difficulty steadily increases:

- AS Chemistry builds from GCSE science
- A2 Chemistry builds from AS Chemistry.

How will you be tested?

Assessment units

For AS Chemistry, you will be tested by three assessment units. For the full A Level in Chemistry, you will take a further three units. AS Chemistry forms 50% of the assessment weighting for the full A Level.

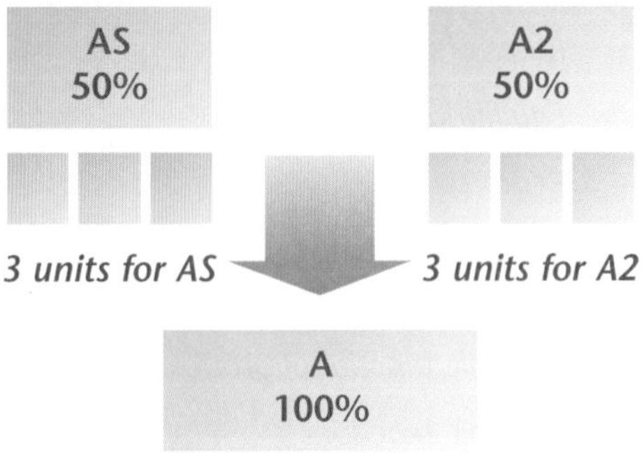

One of the units in AS and in A2 is practically based. You will take two theory units in each of AS and A2. Each theory unit can normally be taken in either January or June. Alternatively, you can study the whole course before taking any of the unit tests.

If you are disappointed with a module result, you can resit each module. The higher mark counts.

Coursework

Coursework forms part of your A Level Chemistry course in the form of the assessment of practical skills. More details are provided on page 7.

Key skills

It is important that you develop your key skills of Communication, Application of Number and Information Technology throughout your AS and A2 courses. These are important skills that you need whatever you do beyond AS and A Levels. To gain the key skills qualification, you will need to demonstrate your ability to put your ideas across to other people, collect data and use up-to-date technology in your work. You will have many opportunities during A2 Chemistry to develop your key skills.

What skills will I need?

For A2 Chemistry, you will be tested by *assessment objectives*: these are the skills and abilities that you should have acquired by studying the course. The assessment objectives for A2 Chemistry are shown below.

Knowledge and understanding of science and How Science Works

- recognising, recalling and showing understanding of scientific knowledge
- selection, organisation and communication of relevant information in a variety of forms.

Application of knowledge and understanding of science and How Science Works

- analysis and evaluation of scientific knowledge and processes
- applying scientific knowledge and processes to unfamiliar situations including those related to issues
- assessing the validity, reliability and credibility of scientific information.

How Science Works

- demonstrating and describing ethical, safe and skilful practical techniques and processes
- selecting appropriate qualitative and quantitative methods
- making, recording and communicating reliable and valid observations and measurements with appropriate precision and accuracy
- analysis, interpretation, explanation and evaluation of the methodology, results and impact of their own and others' experimental and investigative activities in a variety of ways.

Experimental and investigative skills

Chemistry is a practical subject and part of the assessment of A2 Chemistry will test your practical skills. This will be carried out during your lessons. You will be assessed on three main skills:

- planning and implementing
- analysing evidence and drawing conclusions
- evaluating evidence and procedures.

The skills may be assessed in separate practical exercises. They may also be assessed all together in the context of a single 'whole investigation'.

You will receive guidance about how your practical skills will be assessed from your teacher. This Study Guide concentrates on preparing you for the written examinations testing the subject content of A2 Chemistry.

Synoptic assessment

- bringing together knowledge, principles and concepts from different areas of Chemistry and applying them in a particular context
- using chemical skills in contexts which bring together different areas of the subject.

More details of synoptic assessment in Chemistry are discussed in Chapter 6 of this Study Guide (pages 160–163).

Different types of questions in A2 examinations

In A2 Chemistry examinations, different types of question are used to assess your abilities and skills. Unit tests mainly use structured questions requiring both short answers and more extended answers.

Short-answer questions

A short-answer question may test recall or it may test understanding. Short answer questions normally have space for the answers printed on the question paper.

Here are some examples (the answers are shown in blue):

What is meant by isotopes?

Atoms of the same element with different masses.

Calculate the amount (in mol) of H_2O in 4.5 g of H_2O.

1 mol H_2O has a mass of 18 g. $\therefore$ 4.5 g of H_2O contains 4.5/18 = 0.25 mol H_2O.

Structured questions

Structured questions are in several parts. The parts usually have a common context and they often become progressively more difficult and more demanding as you work your way through the question. A structured question may start with simple recall, then test understanding of a familiar or an unfamiliar situation.

Here is an example of a structured question that becomes progressively more demanding.

(a) Write down the atomic structure of the two isotopes of potassium: ^{39}K and ^{41}K.

(i) ^{39}K ...19... protons; ...20... neutrons; ...19... electrons. ✓

(ii) ^{41}K ...19... protons; ...22... neutrons; ...19... electrons. ✓ [2]

(b) A sample of potassium has the following percentage composition by mass: ^{39}K: 92%; ^{41}K: 8% Calculate the relative atomic mass of the potassium sample.

92 x 39/100 + 8 x 41/100 = 39.16 ✓ [1]

(c) What is the electronic configuration of a potassium atom?

$1s^2 2s^2 2p^6 3s^2 3p^6 4s^1$ ✓ [1]

(d) The second ionisation energy of potassium is much larger than its first ionisation energy.

(i) Explain what is meant by the *first ionisation energy* of potassium.

The energy required to remove an electron ✓ from each atom in 1 mole ✓ of gaseous atoms ✓

(ii) Why is there a large difference between the values for the first and the second ionisation energies of potassium?

The 2nd electron removed is from a different shell ✓ which is closer to the nucleus and experiences more attraction from the nucleus. ✓ This outermost electron experiences less shielding from the nucleus because there are fewer inner electron shells than for the 1st ionisation energy. ✓ [6]

Extended answers

In A2 Level Chemistry, questions requiring more extended answers may form part of structured questions or may form separate questions. They may appear anywhere on the paper and will typically have between 5 and 15 marks allocated to the answers as well as several lines of answer space. These questions are also often used to assess your abilities to communicate ideas and put together a logical argument.

The correct answers to extended questions are often less well-defined than to those requiring short answers. Examiners may have a list of points for which credit is awarded up to the maximum for the question.

An example of a question requiring an extended answer is shown below.

> Choose two complex ions of copper with different shapes.
> Describe the shapes of, and bond angles in, these complex ions and explain what is meant by ligand substitution. [6 marks]

> In aqueous condition, Cu^{2+} forms $[Cu(H_2O)_6]^{2+}$ ✓ complex ions. $[Cu(H_2O)_6]^{2+}$ has an octahedral shape and bond angles of 90° ✓. When Cl^- ions are added, a ligand substitution reaction takes place forming $CuCl_4^{2-}$ ✓ complex ions. $CuCl_4^{2-}$ has a tetrahedral shape and bond angles of 109.5°. ✓
>
> Ligand substitution involves one ligand being exchanged by another ligand: ✓
>
> $[Cu(H_2O)_6]^{2+} + 4Cl^- \rightleftharpoons CuCl_4^{2-} + 6H_2O$ ✓

In this type of response, there may be an additional mark for a clear, well-organised answer, using specialist terms. In addition, marks may be allocated for legible text with accurate spelling, punctuation and grammar.

Other types of questions

Free-response and open-ended questions allow you to choose the context and to develop your own ideas.

Multiple-choice or objective questions require you to select the correct response to the question from a number of given alternatives.

Stretch and Challenge

Stretch and Challenge is a concept that is applied to the structured questions in Units 4 and 5 of the exam papers for A2. In principle, it means that the sub-questions become progressively harder so as to challenge more able students and help differentiate between A and A* students.

Stretch and Challenge questions are designed to test a variety of different skills and your understanding of the material. They are likely to test your ability to make appropriate connections between different areas and apply your knowledge in unfamiliar contexts (as opposed to basic recall).

Exam technique

Links from AS

A2 Chemistry builds from the knowledge and understanding acquired after studying AS Chemistry. This Study Guide has been written so that you will be able to tackle A2 Chemistry from an AS Chemistry background.

You should not need to search for important chemistry from AS Chemistry because cross-references have been included as 'Key points from AS' to the AS Chemistry Study Guide: Revise AS.

What are examiners looking for?

Examiners use instructions to help you to decide the length and depth of your answer.

If a question does not seem to make sense, you may have misread it – read it again!

State, define or list

This requires a short, concise answer, often recall of material that can be learnt by rote.

Explain, describe or discuss

Some reasoning or some reference to theory is required, depending on the context.

Outline

This implies a short response, almost a list of sentences or bullet points.

Predict or deduce

You are not expected to answer by recall but by making a connection between pieces of information.

Suggest

You are expected to apply your general knowledge to a 'novel' situation, one which you have not directly studied during the A2 Chemistry course.

Calculate

This is used when a numerical answer is required. You should always use units in quantities and significant figures should be used with care.

Look to see how many significant figures have been used for quantities in the question and give your answer to this degree of accuracy.

If the question uses 3 sig figs, then give your answer to 3 sig figs also.

Some dos and don'ts

Dos

Do answer the question.

- No credit can be given for good Chemistry that is irrelevant to the question.

Do use the mark allocation to guide how much you write.

- Two marks are awarded for two valid points - writing more will rarely gain more credit and could mean wasted time or even contradicting earlier valid points.

Do use diagrams, equations and tables in your responses.

- Even in 'essay-type' questions, these offer an excellent way of communicating chemistry.

Do write legibly.

- An examiner cannot give marks if the answer cannot be read.

Do write using correct spelling and grammar. Structure longer essays carefully.

- Marks are now awarded for the quality of your language in exams.

Don'ts

Don't fill up any blank space on a paper.

- In structured questions, the number of dotted lines should guide the length of your answer.
- If you write too much, you waste time and may not finish the exam paper. You also risk contradicting yourself.

Don't write out the question again.

- This wastes time. The marks are for the answer!

Don't contradict yourself.

- The examiner cannot be expected to choose which answer is intended.

Don't spend too much time on a part that you find difficult.

- You may not have enough time to complete the exam. You can always return to a difficult calculation if you have time at the end of the exam.

What grade do you want?

Everyone would like to improve their grades but you will only manage this with a lot of hard work and determination. You should have a fair idea of your natural ability and likely grade in Chemistry and the hints below offer advice on improving that grade.

For a Grade A

You will need to be a very good all-rounder.

- You must go into every exam knowing the work extremely well.
- You must be able to apply your knowledge to new, unfamiliar situations.
- You need to have practised many, many exam questions so that you are ready for the type of question that will appear.

The exams test all areas of the specification and any weaknesses in your Chemistry will be found out. There must be no holes in your knowledge and understanding. For a Grade A, you must be competent in all areas.

For a Grade C

You must have a reasonable grasp of Chemistry but you may have weaknesses in several areas and you will be unsure of some of the reasons for the Chemistry.

- Many Grade C candidates are just as good at answering questions as Grade A students but holes and weaknesses often show up in just some topics.
- To improve, you will need to master your weaknesses and you must prepare thoroughly for the exam. You must become a better all-rounder.

For a Grade E

You cannot afford to miss the easy marks. Even if you find Chemistry difficult to understand and would be happy with a Grade E, there are plenty of questions in which you can gain marks.

- You must memorise all definitions.
- You must practise exam questions to give yourself confidence that you do know some Chemistry. In exams, answer the parts of questions that you know first. You must not waste time on the difficult parts. You can always go back to these later.
- The areas of Chemistry that you find most difficult are going to be hard to score on in exams. Even in the difficult questions, there are still marks to be gained. Show your working in calculations because credit is given for a sound method. You can always gain some marks if you get part of the way towards the solution.

What marks do you need?

The table below shows how your average mark is transferred into a grade.

average	*80%*	*70%*	*60%*	*50%*	*40%*
grade	*A*	*B*	*C*	*D*	*E*

The new A* grade

OCR and Salters have now introduced an A* grade for A level. This follows on from the introduction of this grade at GCSE and is intended to be awarded to a relatively small number of the highest scoring candidates.

To achieve an A* grade you must score over 80% on all six units combined (grade A) but also 90% on the three A2 units combined. This is quite a difficult target!

Four steps to successful revision

Step 1: Understand

- Study the topic to be learned slowly. Make sure you understand the logic or important concepts.
- Mark up the text if necessary – underline, highlight and make notes.
- Re-read each paragraph slowly.

GO TO STEP 2

Step 2: Summarise

- Now make your own revision note summary:
  What is the main idea, theme or concept to be learned?
  What are the main points? How does the logic develop?
  Ask questions: Why? How? What next?
- Use bullet points, mind maps, patterned notes.
- Link ideas with mnemonics, mind maps, crazy stories.
- Note the title and date of the revision notes
  (e.g. Chemistry: Reaction rates, 3rd March).
- Organise your notes carefully and keep them in a file.

This is now in **short-term memory**. You will forget 80% of it if you do not go to Step 3.
GO TO STEP 3, but first take a 10 minute break.

Step 3: Memorise

- Take 25 minute learning 'bites' with 5 minute breaks.
- After each 5 minute break test yourself:
  Cover the original revision note summary
  Write down the main points
  Speak out loud (record on tape)
  Tell someone else
  Repeat many times.

The material is well on its way to **long-term memory**.
You will forget 40% if you do not do step 4. **GO TO STEP 4**

Step 4: Track/Review

- Create a Revision Diary (one A4 page per day).
- Make a revision plan for the topic, e.g. 1 day later, 1 week later, 1 month later.
- Record your revision in your Revision Diary, e.g.
  Chemistry: Reaction rates, 3rd March 25 minutes
  Chemistry: Reaction rates, 5th March 15 minutes
  Chemistry: Reaction rates, 3rd April 15 minutes
  ... and then at monthly intervals.

Chapter 1

Rates and equilibria

The following topics are covered in this chapter:

- Orders and the rate equation
- Determination of reaction mechanisms
- The equilibrium constant, K_c
- Acids and bases
- The pH scale
- pH changes

1.1 Orders and the rate equation

After studying this section you should be able to:

LEARNING SUMMARY

- explain the terms 'rate of reaction', 'order', 'rate constant', 'rate equation'
- construct a rate equation and calculate its rate constant
- use the initial rates method to deduce orders
- use graphs to deduce reaction rates and orders
- understand that a temperature change increases the rate constant
- determine activation energy graphically

Key points from AS

- **Reaction rates**
  Revise AS pages 83–86

During the study of reaction rates in AS Chemistry, reaction rates were described in terms of activation energy and the Boltzmann distribution. For A2 Chemistry, you will build upon this knowledge and understanding by measuring and calculating reaction rates using rate equations.

Orders and the rate equation

OCR M5
SALTERS M4

The rate of a reaction is determined from experimental results. Reaction rate **cannot** be determined from a chemical equation.

Suitable experimental techniques used to obtain rate data for a given reaction include:

- titration
- pH measurement
- colorimetry
- measuring volumes of gases evolved
- measuring mass changes.

The rate of a reaction is usually determined by the **change in concentration** of a reaction species **with time**.

- Units of rate = mol dm^{-3} s^{-1}

  (mol dm^{-3} → concentration; s^{-1} → per time)

Key points from AS

- **What is a reaction rate?**
  Revise AS pages 83–84

Orders of reaction

For a reaction: **A** + **B** + **C** ⟶ products,

- the **order** of the reaction shows how the reaction rate is affected by the concentrations of **A**, **B** and **C**.

[A] means the concentration of reactant A in mol dm^{-3}.

If the order is 0 (zero order) with respect to reactant **A**,

- the rate is unaffected by changes in concentration of **A**:
  rate $\propto$ $[A]^0$

If the order is 1 (first order) with respect to a reactant **B**,

- the rate is doubled by doubling of the concentration of reactant **B**:
  rate $\propto$ $[B]^1$

If the order is 2 (second order) with respect to a reactant **C**,

- the rate is quadrupled by doubling the concentration of reactant **C**:
  $rate \propto [\mathbf{C}]^2$

Combining the information above,

$rate \propto [\mathbf{A}]^0[\mathbf{B}]^1[\mathbf{C}]^2$

$\therefore\ rate = k[\mathbf{B}]^1[\mathbf{C}]^2$

- This expression is called the **rate equation** for the reaction.
- k is the rate constant of the reaction.

A number raised to the power of zero = 1 and zero order species can be omitted from the rate equation.

The rate equation

The rate of a reaction is determined from experimental results.

It **cannot** be determined from the chemical equation.

The rate equation of a reaction shows how the rate is affected by the concentration of reactants. A rate equation can only be determined from experiments.

In general, for a reaction: **A** + **B** ⟶ **C**,
the reaction rate is given by: $rate = k[\mathbf{A}]^m[\mathbf{B}]^n$

- m and n are the **orders of reaction** with respect to **A** and **B** respectively.
- The **overall order** of reaction is $m + n$.
- The reaction rate is measured as the change in concentration of a reaction species with time.
- The units of rate are mol dm^{-3} s^{-1}.

The rate constant k indicates the rate of the reaction:

- a large value of k ⟶ fast rate of reaction
- a small value of k ⟶ slow rate of reaction.

Measuring rates using graphs

OCR M5
SALTERS M4

Concentration/time graphs

It is often possible to measure the concentration of a reactant or product continuously at various times during the course of an experiment.

- From the results, a concentration/time graph can be plotted.
- The shape of this graph can indicate the order of the reaction by measuring the **half-life** of a reactant.

Key points from AS

- **What is a reaction rate?**
  Revise AS pages 83–84

KEY POINT

The half-life of a reactant is the time taken for its concentration to reduce by half.

A first-order reaction has a constant half-life.

Example of a first order graph

The reaction between bromine and methanoic acid is shown below.

$Br_2(aq) + HCOOH(aq) \longrightarrow 2Br^-(aq) + 2H^+(aq) + CO_2(g)$

- During the course of the reaction, the orange bromine colour disappears as Br_2 reacts to form colourless Br^- ions. The order with respect to bromine can be determined by monitoring the rate of disappearance of bromine using a colorimeter.
- The concentration of the other reactant, methanoic acid, is kept virtually constant by using an excess of methanoic acid.

The concentration/time graph from this reaction is shown on the next page. Notice that the **half-life is constant** at 200 s, showing that this reaction is first order with respect to $Br_2(aq)$.

The half-life curve of a first-order reaction is concentration independent.

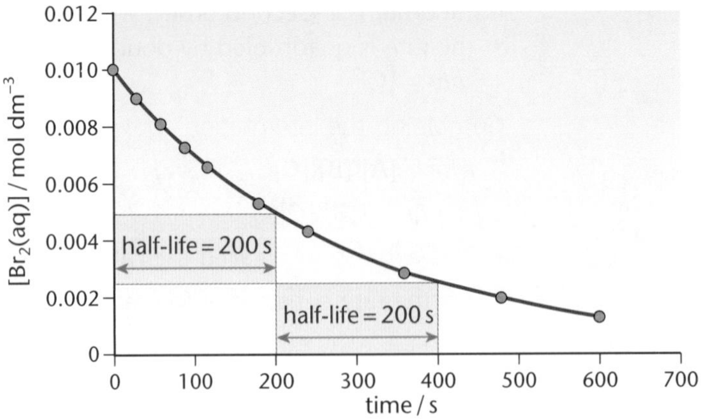

The shapes of concentration/time graphs for zero, first and second order reactions are shown below.

zero order

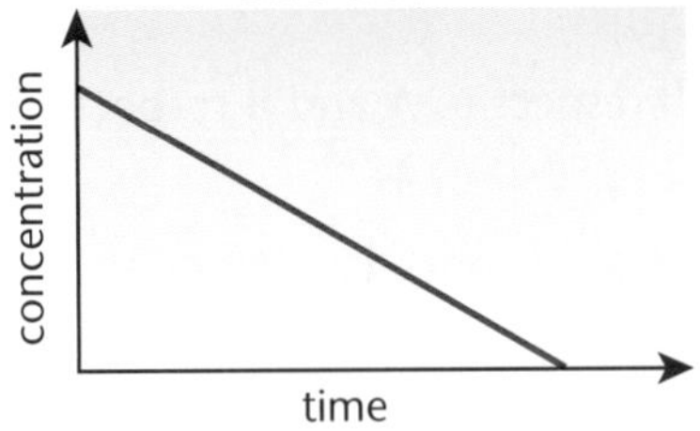

The concentration falls at a steady rate with time

- half-life **decreases** with time.

first order

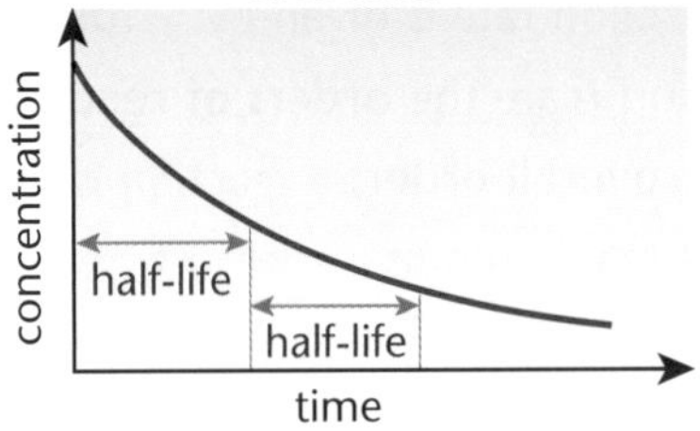

The concentration halves in equal time intervals

- **constant** half-life.

second order

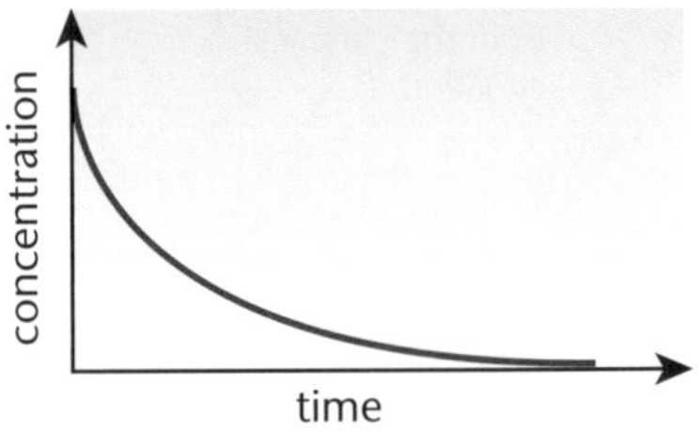

The half-life gets progressively longer as the reaction proceeds

- half-life **increases** with time.

- The first-order relationship can be confirmed by plotting a graph of log[**X**] against time, which gives a straight line.

Measuring rates from tangents

The gradient of the concentration/time graph is a measure of the rate of a reaction.

For the reaction: **A** $\longrightarrow$ **B**, the graph below shows how the concentration of **A** changes during the course of the reaction.

> At any instant of time during the reaction, the rate can be measured by drawing a tangent to the curve.
>
> **KEY POINT**

The initial rate of the reaction is the tangent of a concentration/time graph at time=0.

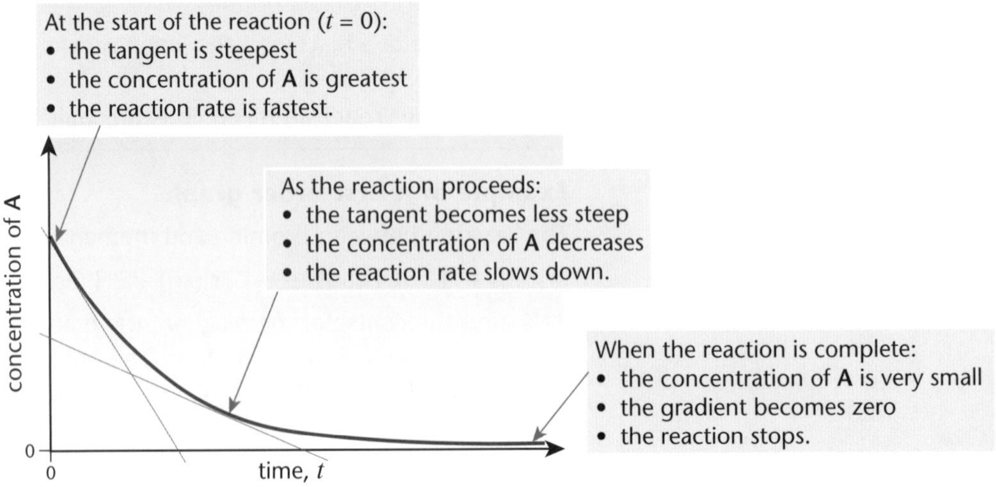

At an instant of time, t, during the reaction:

- the rate of **change** in concentration of **A** = the gradient of the tangent.

The negative sign shows that the concentration of **A** decreases during the reaction.

The rate could also be followed by measuring the rate of **increase** in concentration of the product **B**.

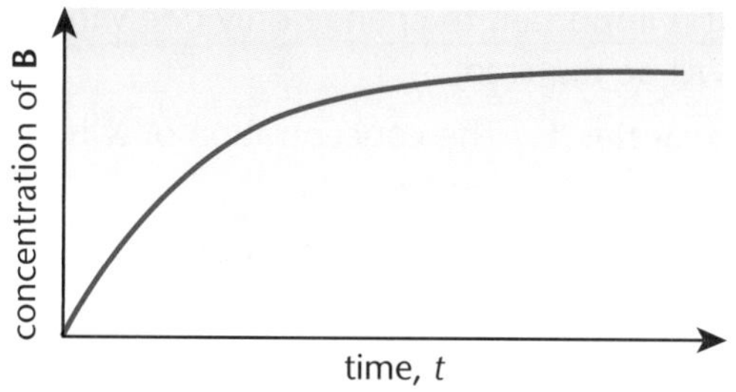

B is formed and its concentration increases during the reaction.

rate of **change** in concentration of **B** = the gradient of the tangent

Plotting a rate/concentration graph

- A concentration/time graph is first plotted.
- **Tangents** are drawn at several time values on the concentration/time graph, giving values of reaction rates.
- A second graph is plotted of **rate** against **concentration**.
- The shape of this graph confirms the order of the reaction.

zero order

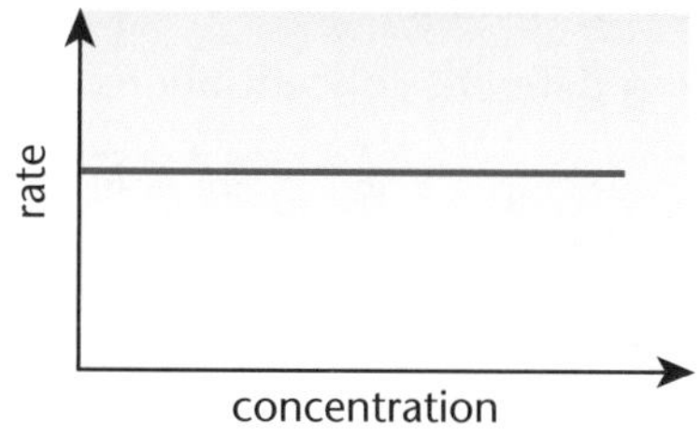

Rate ∝ $[\mathbf{A}]^0$ ∴ rate = constant
- Rate unaffected by changes in concentration.

first order

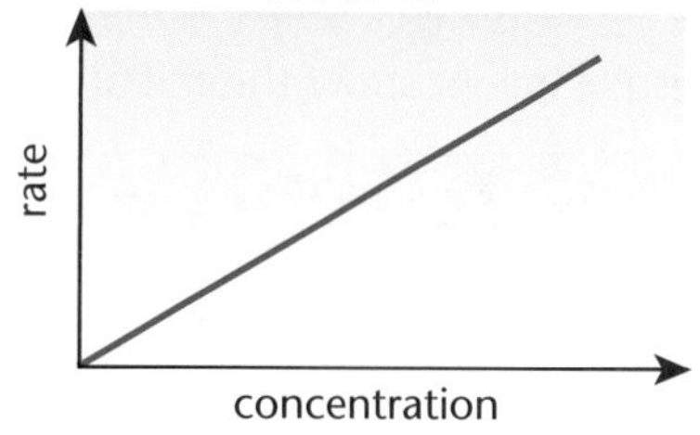

Rate ∝ $[\mathbf{A}]^1$
- Rate doubles as concentration doubles.

second order

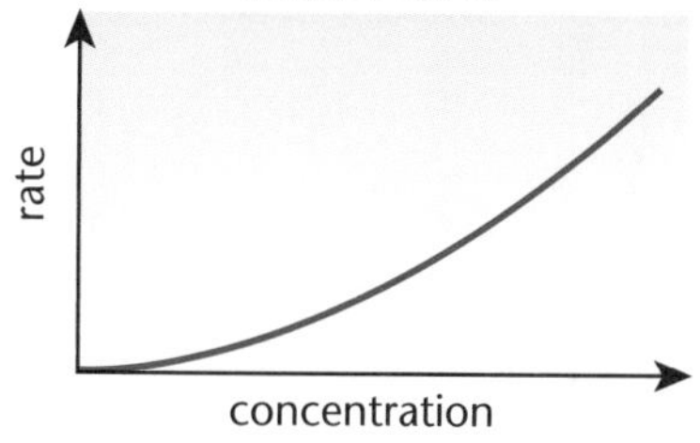

Rate ∝ $[\mathbf{A}]^2$
- Rate quadruples as concentration doubles.

- The second-order relationship can be confirmed by plotting a graph of rate against $[\mathbf{A}]^2$, which gives a straight line.

Progress check

1 The data below shows how the concentration of reactant **A** changes during the course of a reaction.

Time / 10^4 s	0	0.36	0.72	1.08	1.44
[A] /mol dm^{-3}	0.240	0.156	0.104	0.068	0.045

(a) On graph paper, plot a concentration/time graph and, by drawing tangents, estimate:

(i) the initial rate

(ii) the rate 1×10^4 s after the reaction has started.

(b) Determine the half-life of the reaction and show that the reaction is first order with respect to **A**.

1 (a) (i) 2.5×10^{-5} mol dm^{-3} s^{-1} to 3.5×10^{-5} mol dm^{-3} s^{-1} (ii) 8.7×10^{-6} mol dm^{-3} s^{-1}

(b) Half-life = 5.8×10^3 s. Successive half-lives are constant.

Determination of a rate of equation using the initial rates method

OCR M5
SALTERS M4

For a reaction, X + Y ⟶ Z, a series of experiments can be carried out using different **initial** concentrations of the reactants **X** and **Y**.

It is important to change only one variable at a time, so two series of experiments will be required.

- **Series 1** – The concentration of **X is changed** whilst the concentration of **Y is kept constant.**
- **Series 2** – The concentration of **Y is changed** whilst the concentration of **X is kept constant.**

Clock reactions give an approximation to the initial rates method.

Clock reactions are often used to obtain a value for the initial rate of a reaction. A clock reaction measures the **time**, *t*, from the start of the reaction until a visual change – usually a change in colour, appearance of a precipitate or a solid disappearing.

- The clock reaction is repeated several times whilst varying the concentration of one of the reactants. The concentration of any other reactant is kept constant.
- The **initial rate** of the reaction is approximately proportional to **1/*t***.
- A **graph** is then plotted of **initial rate (1/*t*)** against **concentration**.
- The **order** with respect to the reactant of which the concentration has been varied, is found from the **shape of the graph**. (See page 17).

A series of clock reactions is repeated for different concentrations of each reactant. A rate/concentration graph is plotted for each reactant.

The results below show the initial rates using different concentrations of **X** and **Y**.

Experiment	[**X**(aq)] /mol dm^{-3}	[**Y**(aq)] /mol dm^{-3}	initial rate /mol dm^{-3} s^{-1}
1	1.0×10^{-2}	1.0×10^{-2}	0.5×10^{-3}
2	2.0×10^{-2}	1.0×10^{-2}	2.0×10^{-3}
3	2.0×10^{-2}	2.0×10^{-2}	4.0×10^{-3}

Determine the orders

Comparing experiments 1 and 2: [**Y**(aq)] has been kept constant

- [**X**(aq)] has been doubled, rate × 4
  ∴ order with respect to **X**(aq) = 2.

Comparing experiments 2 and 3: [**X**(aq)] has been kept constant

- [**Y**(aq)] has been doubled, rate doubles
  ∴ order with respect to **Y**(aq) = 1.

Use the orders to write the rate equation

- $rate = k[\mathbf{X}(aq)]^2[\mathbf{Y}(aq)]$
- The overall order of this reaction is (2 + 1) = third order

Calculate the rate constant for the reaction

Rearrange the rate equation:

$$\text{The rate constant, } k = \frac{rate}{[\mathbf{X}(aq)]^2[\mathbf{Y}(aq)]}$$

Calculate k using values from one of the experiments.

$$k = \frac{(2.0 \times 10^{-3})}{(2.0 \times 10^{-2})^2\,(1.0 \times 10^{-2})} = 500 \text{ dm}^6 \text{ mol}^{-2} \text{ s}^{-1}$$

In this example, the results from Experiment 2 have been used.

You will get the same value of *k* from the results of **any** of the experimental runs.

Try calculating *k* from Experiment 1 and from Experiment 3. All give the same value.

Units of rate constants

Notice how units need to be worked out afresh for reactions with different overall orders.

You do not need to memorise these but it is important that you are able to work these out when needed.

See also page 24 in which the units of the equilibrium constant K_c are discussed.

The units of a rate constant depend upon the rate equation for the reaction. We can determine units of k by substituting units for rate and concentration into the rearranged rate equation. The table below shows how units can be determined.

For a **zero-order** reaction: $rate = k[\mathbf{A}]^0 = k$ units of k = **mol dm^{-3} s^{-1}**	For a **first-order** reaction: $rate = k[\mathbf{A}] \quad \therefore\ k = \frac{rate}{[\mathbf{A}]}$ Units of $k = \frac{(\text{mol dm}^{-3}\ \text{s}^{-1})}{(\text{mol dm}^{-3})}$ = **s^{-1}**
For a **second-order** reaction: $rate = k[\mathbf{A}]^2 \quad \therefore\ k = \frac{rate}{[\mathbf{A}]^2}$ Units of $k = \frac{(\text{mol dm}^{-3}\ \text{s}^{-1})}{(\text{mol dm}^{-3})^2}$ = **dm^3 mol^{-1} s^{-1}**	For a **third-order** reaction: $rate = k[\mathbf{A}]^2[\mathbf{B}] \quad \therefore\ k = \frac{rate}{[\mathbf{A}]^2[\mathbf{B}]}$ Units of $k = \frac{(\text{mol dm}^{-3}\ \text{s}^{-1})}{(\text{mol dm}^{-3})^2\ (\text{mol dm}^{-3})}$ = **dm^6 mol^{-2} s^{-1}**

Progress check

1 A chemical reaction is first order with respect to compound **P** and second order with respect to compound **Q**.

(a) Write the rate equation for this reaction.

(b) What is the overall order of this reaction?

(c) By what factor will the rate increase if:

(i) the concentration of **P only** is doubled

(ii) the concentration of **Q only** is doubled

(iii) the concentrations of **P** and **Q** are **both** doubled?

(d) What are the units of the rate constant of this reaction?

2 The reaction of ethanoic anhydride, $(CH_3CO)_2O$, with ethanol, C_2H_5OH, can be represented by the equation:

$$(CH_3CO)_2O + C_2H_5OH \longrightarrow CH_3CO_2C_2H_5 + CH_3CO_2H$$

The table below shows the initial concentrations of the two reactants and the initial rates of reaction.

Experiment	*$[(CH_3CO)_2O]$ /mol dm^{-3}*	*$[C_2H_5OH]$ /mol dm^{-3}*	*initial rate /mol dm^{-3} s^{-1}*
1	0.400	0.200	6.60×10^{-4}
2	0.400	0.400	1.32×10^{-3}
3	0.800	0.400	2.64×10^{-3}

(a) State and explain the order of reaction with respect to:

(i) ethanoic anhydride (ii) ethanol.

(b) (i) Write an expression for the overall rate equation.

(ii) What is the overall order of reaction?

(c) Calculate the value, with units, for the rate constant, k.

1 (a) $rate = k[\mathbf{P}][\mathbf{Q}]^2$

(b) third order

(c) (i) rate doubles (ii) rate quadruples (iii) rate × 8

(d) dm^6 mol^{-2} s^{-1}

2 (a) (i) first order; concentration doubles; rate doubles

(ii) first order; concentration doubles; rate doubles.

(b) (i) $rate = k\,[(CH_3CO)_2O]\,[C_2H_5OH]$ (ii) second order.

(c) 8.25×10^{-3} dm^3 mol^{-1} s^{-1}.

How does the rate constant, k, vary with temperature?

OCR M5
SALTERS M4

Key points from AS

- **Boltzmann distribution**
- **Activation energy**

Revise AS page 84

During AS Chemistry, you studied how the rate of reaction increases with temperature in terms of collisions, the activation energy of the reaction and the Boltzmann distribution of molecular energies.

The effect of temperature on rate constants

An increase in temperature speeds up the rate of most reactions and the rate constant, *k*, increases.

The table below shows the increase in the rate constant for the decomposition of hydrogen iodide with increasing temperature.

$$2HI(g) \longrightarrow H_2(g) + I_2(g)$$

For many reactions, the rate doubles for each 10°C increase in temperature.

temperature/K	556	629	700	781
rate constant, k /dm^3 mol^{-1} s^{-1}	7.04×10^{-7}	6.04×10^{-5}	2.32×10^{-3}	7.90×10^{-2}

A reaction with a large activation energy has a small rate constant. Such a reaction may need a considerable temperature rise to increase the value of the rate constant sufficiently for a reaction to take place at all.

Why are catalysts so effective?

A catalyst is so effective because:

- it **reduces** the **activation energy**
- which **increases** the **rate of reaction**
- which **increases** the **rate constant**, *k*.

1.2 Determination of reaction mechanisms

After studying this section you should be able to:

- *understand what is meant by a rate-determining step*
- *predict a rate equation from a rate-determining step*
- *predict a rate-determining step from a rate equation*
- *use a rate equation and the overall equation for a reaction to predict a possible reaction mechanism*

LEARNING SUMMARY

Predicting reaction mechanisms

OCR M5
SALTERS M4

Key points from AS

- **The hydrolysis of halogenoalkanes** *Revise AS pages 116–117*

The rate-determining step

Chemical reactions often take place in a series of steps. The detail of these steps is the **reaction mechanism**.

> The rate equation can provide clues about a likely reaction mechanism by identifying the **slowest** stage of a reaction sequence, called the **rate-determining step**.
>
> KEY POINT

You are **not** expected to remember these examples.

You **are** expected to interpret data to identify a rate-determining step and to suggest possible steps in the mechanism for the reaction.

The hydrolysis of tertiary halogenoalkanes

2-bromo-2-methylpropane, $(CH_3)_3C{-}Br$, is hydrolysed by aqueous alkali, OH^-.

Experiments show that the rate equation for this reaction is:

$$rate = k[(CH_3)_3C{-}Br]$$

- The rate equation shows that a **slow** reaction step must involve **only** $(CH_3)_3C{-}Br$.
- OH^- has **no effect** on the reaction rate.

This supports the slow step below:

$$\underbrace{(CH_3)_3C{-}Br}_{\text{One particle in rate-determining step}} \longrightarrow (CH_3)_3C^+ + Br^- \qquad \textbf{SLOW}$$

This reaction proceeds via a two-step mechanism: **1** particle, $(CH_3)_3CBr$, is involved in the rate-determining step of a **N**ucleophilic **S**ubstitution mechanism.

The overall equation for the hydrolysis of 1-bromobutane by aqueous alkali is shown below:

$$(CH_3)_3C{-}Br + OH^- \longrightarrow (CH_3)_3C{-}OH + Br^-$$

- This **does not** match the equation for the slow step above.
- The rate-determining step must be followed by further **fast** steps.
- **OH^-** must be involved as it is in the **overall equation**.

A possible second step is shown below:

$$(CH_3)_3C^+ + OH^- \longrightarrow (CH_3)_3C{-}OH \qquad \textbf{FAST}$$

The **overall equation** is the sum of the equations from each step in the mechanism.

In the hydrolysis of a tertiary halogenoalkane:

1st step	$(CH_3)_3C{-}Br$	$\longrightarrow$	$(CH_3)_3C^+ + Br^-$	**SLOW**
2nd step	$(CH_3)_3C^+ + OH^-$	$\longrightarrow$	$(CH_3)_3C{-}OH$	**FAST**
Overall equation	$(CH_3)_3C{-}Br + OH^-$	$\longrightarrow$	$(CH_3)_3C{-}OH + Br^-$	

Notice that $(CH_3)_3C^+$ does not feature in the overall equation. It is generated in one step and used up in another step.

The hydrolysis of primary halogenoalkanes

1-Bromobutane, $CH_3CH_2CH_2CH_2Br$, is hydrolysed by aqueous alkali, OH^-.

Experiments show that the rate equation for this reaction is:

rate = $k[CH_3CH_2CH_2CH_2Br]\,[OH^-]$

The **slow** reaction step must involve **both** $CH_3CH_2CH_2CH_2Br$ **and** OH^-.

This reaction proceeds via a one-step mechanism:

2 particles, $CH_3CH_2CH_2CH_2Br$ **and** OH^-, are involved in the rate-determining step of a **N**ucleophilic **S**ubstitution mechanism.

This supports the slow step below:

$$\underbrace{CH_3CH_2CH_2CH_2Br + OH^-}_{\textbf{Two}\text{ particles in rate-determining step}} \longrightarrow CH_3CH_2CH_2CH_2OH + Br^- \qquad \textbf{SLOW}$$

The overall equation for the hydrolysis of 1-bromobutane by aqueous alkali is:

$$CH_3CH_2CH_2CH_2Br + OH^- \longrightarrow CH_3CH_2CH_2CH_2OH + Br^-$$

- This **does** match the equation for the slow step above.
- The reaction **mechanism** must have just a **single step**.

The reaction of iodine and propanone in acid

The reaction of iodine, I_2, and propanone, CH_3COCH_3, is catalysed by acid, H^+.

The overall equation for the acid-catalysed reaction of iodine and propanone is:

$$I_2 + CH_3COCH_3 \longrightarrow CH_3COCH_2I + H^+ + I^- \qquad \textit{overall equation}$$

The slow step is the **rate-determining step**.

Experiments show that the rate equation for this reaction is:

rate = $k[CH_3COCH_3]\,[H^+]$

If the species in the rate equation do not match those in the overall equation, the reaction mechanism must have more than one step.

The **slow** rate-determining step must involve **both** CH_3COCH_3 **and** H^+.

A possible equation for the rate-determining step is shown below:

$$\underbrace{CH_3COCH_3 + H^+}_{\textbf{Two}\text{ particles in rate-determining step}} \longrightarrow CH_3C(OH^+)CH_3 \qquad \textbf{SLOW}$$

- This does **not** match the overall equation.

Possible further steps are shown below:

$$CH_3C(OH^+)CH_3 \longrightarrow CH_2{=}C(OH)CH_3 + H^+ \qquad \textbf{FAST}$$

$$CH_3C(OH){=}CH_2 + I_2 \longrightarrow CH_3(OH)CICH_2I \qquad \textbf{FAST}$$

$$CH_3(OH)CICH_2I \longrightarrow CH_3COCH_2I + H^+ + I^- \qquad \textbf{FAST}$$

H^+ is a catalyst.

It is needed for the first step but is regenerated in a further step.

Overall, it is not consumed.

KEY POINT

The overall equation is the sum of the equations from each step in the mechanism.

- The rate of the reaction is controlled mainly by the slowest step of the mechanism – the **rate-determining step**.
- The rate equation only includes reacting species involved in the slow rate-determining step.
- The orders in the rate equation match the number of species involved in the rate-determining step.

Progress check

1 CH_3CH_2Br and cyanide ions, CN^- react as follows:

$$CH_3CH_2Br + CN^- \longrightarrow CH_3CH_2CN + Br^-$$

Show two different possible reaction routes for this reaction and write down the expected rate equation for each route.

1 Single step: $CH_3CH_2Br + CN^- \longrightarrow CH_3CH_2CN + Br^-$ *rate* $= k[CH_3CH_2Br][CN^-]$

Two steps: $CH_3CH_2Br \longrightarrow CH_3CH_2^+ + Br^-$ slow

$CH_3CH_2^+ + CN^- \longrightarrow CH_3CH_2CN$ fast *rate* $= k[CH_3CH_2Br]$

1.3 The equilibrium constant, K_c

After studying this section you should be able to:

- *deduce, for homogeneous reactions, K_c in terms of concentrations*
- *calculate values of K_c given appropriate data*
- *understand that K_c is changed only by changes in temperature*
- *understand how the magnitude of K_c relates to the equilibrium position*
- *calculate, from data, the concentrations present at equilibrium*

LEARNING SUMMARY

Key points from AS

- **Chemical equilibrium**
  Revise AS pages 88–93

During the study of equilibrium in AS Chemistry, the concept of dynamic equilibrium is introduced and le Chatelier's principle is used to predict how a change in conditions may alter the equilibrium position. For A2 Chemistry, you will build upon this knowledge and understanding to find out the exact position of equilibrium using the Equilibrium Law.

The equilibrium law

OCR M5
SALTERS M5

The equilibrium law states that, for an equation:

$$a\,\mathbf{A} + b\,\mathbf{B} \rightleftharpoons c\,\mathbf{C} + d\,\mathbf{D},$$

$$K_c = \frac{[\mathbf{C}]^c\,[\mathbf{D}]^d}{[\mathbf{A}]^a\,[\mathbf{B}]^b}$$

- $[\mathbf{A}]^a$, etc., are the **equilibrium** concentrations of the reactants and products of the reaction.
- Each product and reactant has its equilibrium concentration raised to the **power** of its **balancing number** in the equation.
- The equilibrium concentrations of the **products** are multiplied together on **top** of the fraction.
- The equilibrium concentrations of the **reactants** are multiplied together on the **bottom** of the fraction.

The overall equation for the reaction is used to determine K_c.

Working out K_c

For the equilibrium: $H_2(g) + I_2(g) \rightleftharpoons 2HI(g)$

applying the equilibrium law above: $K_c = \dfrac{[HI(g)]^2}{[H_2(g)]\,[I_2(g)]}$

Equilibrium concentrations of $H_2(g)$, $I_2(g)$ and $HI(g)$ are shown below:

$[H_2(g)]$ /mol dm^{-3}	$[I_2(g)]$ /mol dm^{-3}	$[HI(g)]$ /mol dm^{-3}
0.0114	0.00120	0.0252

$$K_c = \frac{[HI(g)]^2}{[H_2(g)]\,[I_2(g)]} = \frac{0.0252^2}{0.0114 \times 0.00120} = 46.4$$

In exams you are likely to be assessed only on **homogeneous equilibria** – all species are in the same phase:
all gaseous (g)
or
all aqueous (aq)
or
all liquid (l).

Units of K_c

- In the K_c expression, each concentration value is replaced by its units:

$$K_c = \frac{[HI(g)]^2}{[H_2(g)]\,[I_2(g)]} = \frac{(\text{mol dm}^{-3})^2}{(\text{mol dm}^{-3})\,(\text{mol dm}^{-3})}$$

- For this equilibrium, the units cancel.

∴ in the equilibrium: $H_2(g) + I_2(g) \rightleftharpoons 2HI(g)$, K_c has no units.

The units must be worked out afresh for each equilibrium.
See also page 19 in which the units of the rate constant are discussed.

Properties of K_c

The magnitude of K_c indicates the extent of a chemical equilibrium.

- $K_c = 1$ indicates an equilibrium halfway between reactants and products.
- $K_c = 100$ indicates an equilibrium well in favour of the products.
- $K_c = 1 \times 10^{-2}$ indicates an equilibrium well in favour of the reactants.

K_c indicates how FAR a reaction proceeds; **not** how FAST.
K_c is unaffected by changes in concentration or pressure.
K_c can **only** be changed by altering the temperature.

Although changes in K_c affect the equilibrium yield, the actual conditions used may be different.

If the forward reaction is **exothermic**, K_c increases and the equilibrium yield increases as temperature decreases.

However, a decrease in temperature may slow down the reaction so much that the reaction is stopped.

In practice, a compromise will be needed where equilibrium and rate are considered together – a reasonable equilibrium yield must be obtained in a reasonable length of time.

How does K_c vary with temperature?

It cannot be stressed too strongly that K_c can be changed only by altering the temperature. How K_c changes depends upon whether the reaction gives out or takes in heat energy.

If the forward reaction is **exothermic**, K_c **decreases** with increasing temperature. Raising the temperature reduces the equilibrium yield of products.

$$H_2(g) + I_2(g) \rightleftharpoons 2HI(g): \quad \Delta H^{\ominus}_{298} = -9.6 \text{ kJ mol}^{-1}$$

temperature /K	500	700	1100
K_c	160	54	25

If the forward reaction is **endothermic**, K_c **increases** with increasing temperature. Raising the temperature increases the equilibrium yield of products.

$$N_2(g) + O_2(g) \rightleftharpoons 2NO(g): \quad \Delta H^{\ominus}_{298} = +180 \text{ kJ mol}^{-1}$$

temperature /K	500	700	1100
K_c	5×10^{-13}	4×10^{-8}	1×10^{-5}

Progress check

1 For each of the following equilibria, write down the expression for K_c. State the units for K_c for each reaction.

(a) $N_2O_4(g) \rightleftharpoons 2NO_2(g)$
(b) $CO(g) + 2H_2(g) \rightleftharpoons CH_3OH(g)$
(c) $H_2(g) + Br_2(g) \rightleftharpoons 2HBr(g)$
(d) $2SO_2(g) + O_2(g) \rightleftharpoons 2SO_3(g)$

2 Explain whether the two reactions, **A** and **B**, are exothermic or endothermic.

temperature /K	*numerical value of K_c*	
	reaction A	***reaction B***
200	5.51×10^{-8}	4.39×10^{4}
400	1.46	4.03
600	3.62×10^{2}	3.00×10^{-2}

1 (a) $K_c = \frac{[NO_2(g)]^2}{[N_2O_4(g)]}$ mol dm^{-3} (b) $K_c = \frac{[CH_3OH(g)]}{[CO(g)][H_2(g)]^2}$ dm^6 mol^{-2}

(c) $K_c = \frac{[HBr(g)]^2}{[H_2(g)][Br_2(g)]}$ (d) $K_c = \frac{[SO_3(g)]^2}{[SO_2(g)]^2[O_2(g)]}$ dm^3 mol^{-1}

2 Reaction **A**: endothermic; K_c increases with increasing temperature.
Reaction **B**: exothermic; K_c decreases with increasing temperature.

Determination of K_c from experiment

OCR MS
SALTERS MS

The equilibrium constant K_c can be calculated using experimental results. The example below shows how to:

- determine the equilibrium concentrations of the components in an equilibrium mixture
- calculate K_c.

The ethyl ethanoate esterification equilibrium

The most important stage in this calculation is to find the **change** in the number of moles of each species in the equilibrium.

0.200 mol CH_3COOH and 0.100 mol C_2H_5OH were mixed together with a trace of acid catalyst in a total volume of 250 cm^3. The mixture was allowed to reach equilibrium:

$$CH_3COOH + C_2H_5OH \rightleftharpoons CH_3COOC_2H_5 + H_2O.$$

Analysis of the mixture showed that 0.115 mol of CH_3COOH were present at equilibrium.

First summarise the results

It is useful to summarise the results as an 'I.C.E.' table: '**I**nitial/**C**hange/**E**quilibrium'.

	CH_3COOH	C_2H_5OH	$CH_3COOC_2H_5$	H_2O
Initial no. of moles	0.200	0.100	0	0
Change in moles				
Equilibrium no. of moles	0.115			

From the CH_3COOH values:

- moles of CH_3COOH that reacted = 0.200 – 0.115 = **0.085** mol

Find the change in moles of each component in the equilibrium

The amount of each component can be determined from the balanced equation. The change in moles of CH_3COOH is already known.

In this experiment, 0.085 mol CH_3COOH has **reacted** with 0.085 mol C_2H_5OH to form 0.085 mol $CH_3COOC_2H_5$ and 0.085 mol H_2O.

equation:	CH_3COOH +	C_2H_5OH	$\rightleftharpoons$	$CH_3COOC_2H_5$ +	H_2O
molar quantities:	1 mol	1 mol	$\longrightarrow$	1 mol	1 mol
change/mol:	**–0.085**	–0.085		+0.085	+0.085

Equilibrium concentrations are determined for each component

Note that the total volume is 250 cm^3 (0.250 dm^3). The concentration must be expressed as mol dm^{-3}.

	CH_3COOH +	C_2H_5OH $\rightleftharpoons$	$CH_3COOC_2H_5$ +	H_2O
Initial amount/mol	0.200	0.100	0	0
Change in moles	–0.085	–0.085	+0.085	+0.085
Equilibrium amount/mol	0.115	0.015	0.085	0.085
Equilibrium conc. /mol dm^{-3}	$\frac{0.115}{0.250}$	$\frac{0.015}{0.250}$	$\frac{0.085}{0.250}$	$\frac{0.085}{0.250}$

Write the expression for K_c and substitute values

$$K_c = \frac{[CH_3COOC_2H_5]\,[H_2O]}{[CH_3COOH]\,[C_2H_5OH]} = \frac{\frac{0.085}{0.250} \times \frac{0.085}{0.250}}{\frac{0.115}{0.250} \times \frac{0.015}{0.250}}$$

Calculate K_c

$\therefore K_c = 4.19$ (no units: all units cancel)

Progress check

1 Several experiments were set up for the $H_2(g)$, $I_2(g)$ and $HI(g)$ equilibrium. The equilibrium concentrations are shown below.

$[H_2(g)]$ /mol dm^{-3}	$[I_2(g)]$ /mol dm^{-3}	$[HI(g)]$ /mol dm^{-3}
0.0092	0.0020	0.0296
0.0077	0.0031	0.0334
0.0092	0.0022	0.0308
0.0035	0.0035	0.0235

Calculate the value for K_c for each experiment. Hence, show that each experiment has the same value for K_c (allowing for experimental error). Work out an average value for K_c.

2 2 moles of ethanoic acid, CH_3COOH were mixed with 3 moles of ethanol and the mixture was allowed to reach equilibrium.

$$CH_3COOH + C_2H_5OH \rightleftharpoons CH_3COOC_2H_5 + H_2O$$

At equilibrium, 0.5 moles of ethanoic acid remained.

(a) Work out the equilibrium concentrations of each component in the mixture. (You will need to use V to represent the volume but this will cancel out in your calculation.)

(b) Use these values to calculate K_c.

3 When 0.50 moles of $H_2(g)$ and 0.18 moles of $I_2(g)$ were heated at 500°C, the equilibrium mixture contained 0.01 moles of $I_2(g)$.

The equation is:

$H_2(g) + I_2(g) \rightleftharpoons 2HI(g)$

(a) How many moles of $I_2(g)$ reacted?

(b) How many moles of $H_2(g)$ were present at equilibrium?

(c) How many moles of $HI(g)$ were present at equilibrium?

(d) Calculate the equilibrium constant, K_c for this reaction.

1 values: 47.6; 46.7; 46.9; 45.1. average = 46.6

2 (a) CH_3COOH, 0.5/V mol dm^{-3}; C_2H_5OH, 1.5/V mol dm^{-3}; $CH_3COOC_2H_5$, 1.5/V mol dm^{-3}; H_2O, 1.5/V mol dm^{-3}.

(b) K_c = 3.

3 (a) 0.17 mol (b) 0.33 mol (c) 0.34 mol (d) 35

1.4 Acids and bases

After studying this section you should be able to:

- *describe what is meant by Brønsted–Lowry acids and bases*
- *understand conjugate acid–base pairs*
- *understand the difference between a strong and a weak acid*
- *define the acid dissociation constant, K_a*

LEARNING SUMMARY

Key points from AS

- **Calculations in acid–base titrations**
  Revise AS page 34
- **Acids and bases**
  Revise AS pages 38–40

During the study of acids and bases at GCSE, you learnt that the pH scale can be used to measure the strength of acids and bases. You also learnt the reactions of acids with metals, carbonates and alkalis. During AS Chemistry, you revisited these reactions and found out how titrations can be used to measure the concentration of an unknown acid or base. For A2 Chemistry, you will study acids in terms of proton transfer, strength, pH and buffers.

Brønsted–Lowry acids and bases

OCR M5
SALTERS M4, M5

A molecule of an acid contains a hydrogen atom that can be released as a positive hydrogen ion or proton, H^+.

- An acid is a substance producing an excess of hydrogen ions in solution – the Arrhenius theory.
- An acid is a proton donor – the Brønsted–Lowry theory.

At A level, we use the Brønsted–Lowry model of acids and bases in terms of proton transfer.

An acid is a proton donor.

A base is a proton acceptor.

- A Brønsted–Lowry **acid** is a proton donor.
- A Brønsted–Lowry **base** is a proton acceptor.
- An **alkali** is a base that dissolves in water forming $OH^-(aq)$ ions.

KEY POINT

Key points from AS

- **Acids, bases and alkalis**
  Revise AS pages 38–39

The role of $H^+(aq)$ in acid reactions (OCR only)

The H^+ ion is the active part in typical acid reactions. The ionic equations below emphasise this role. The reaction is essentially the same, irrespective of the acid involved.

Salts are always formed in solution from these typical acid reactions. The salt can be isolated by evaporation of water.

H_2SO_4 forms sulfates

HCl forms chlorides

HNO_3 forms nitrates

Acids react with carbonates, forming carbon dioxide:

solid carbonate:

$$2H^+(aq) + CaCO_3(s) \longrightarrow Ca^{2+}(aq) + H_2O(l) + CO_2(g)$$

aqueous carbonate:

$$2H^+(aq) + CO_3^{2-}(aq) \longrightarrow H_2O(l) + CO_2(g)$$

Acids react with bases (metal oxides), forming water:

$$2H^+(aq) + CaO(s) \longrightarrow Ca^{2+}(aq) + H_2O(l)$$

Acids react with alkalis, forming water:

$$H^+(aq) + OH^-(aq) \longrightarrow H_2O(l)$$

Acids react with many metals, forming hydrogen:

$$2H^+(aq) + Mg(s) \longrightarrow Mg^{2+}(aq) + H_2(g)$$

With most metals, nitric acid reacts in a different way. If you are asked for the reaction of an acid with a metal, choose a different acid!

Acid–base pairs

Acids and bases are linked by H^+ as **conjugate pairs:**

- the **conjugate acid** donates H^+
- the **conjugate base** accepts H^+.

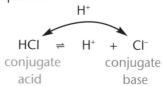

H^+

$$HCl \rightleftharpoons H^+ + Cl^-$$

conjugate acid — conjugate base

Examples of some conjugate acid–base pairs are shown below.

	acid				base
hydrochloric acid	HCl	$\rightleftharpoons$	H^+	+	Cl^-
sulfuric acid	H_2SO_4	$\rightleftharpoons$	H^+	+	HSO_4^-
ethanoic acid	CH_3COOH	$\rightleftharpoons$	H^+	+	CH_3COO^-

An acid needs a base

An acid can only donate a proton if there is a base to accept it. Most reactions of acids take place in aqueous conditions with water acting as the base. By mixing an acid with a base, an equilibrium is set up comprising **two** acid–base **conjugate pairs**.

The equilibrium system in aqueous ethanoic acid is shown below.

You should be able to identify acid–base pairs in equations such as this.

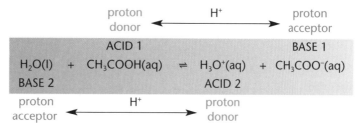

proton donor ← H^+ → proton acceptor

ACID 1 — BASE 1

$$H_2O(l) + CH_3COOH(aq) \rightleftharpoons H_3O^+(aq) + CH_3COO^-(aq)$$

BASE 2 — ACID 2

proton acceptor ← H^+ → proton donor

In equations, the oxonium ion H_3O^+ is usually shown as $H^+(aq)$.

- In a reaction involving an aqueous acid such as hydrochloric acid, the 'active' species is the **oxonium** ion, H_3O^+ (ACID 2 above).
- Formation of the oxonium ion requires **both** an acid **and** water.

Progress check

1 Identify the acid–base pairs in the acid–base equilibria below.
(a) $HNO_3 + H_2O \rightleftharpoons H_3O^+ + NO_3^-$
(b) $NH_3 + H_2O \rightleftharpoons NH_4^+ + OH^-$
(c) $H_2SO_4 + H_2O \rightleftharpoons HSO_4^- + H_3O^+$

2 Write equations for the following acid–base equilibria:
(a) hydrochloric acid and hydroxide ions
(b) ethanoic acid, CH_3COOH, and water.

1 (a) acid 1: HNO_3, base 1: NO_3^-; acid 2: H_3O^+, base 2: H_2O
(b) acid 1: H_2O, base 1: OH^-; acid 2: NH_4^+, base 2: NH_3
(c) acid 1: H_2SO_4, base 1: HSO_4^-; acid 2: H_3O^+, base 2: H_2O
2 (a) $HCl + OH^- \rightleftharpoons H_2O + Cl^-$
(b) $CH_3COOH + H_2O \rightleftharpoons H_3O^+ + CH_3COO^-$

Strength of acids and bases

OCR M5
SALTERS M5

The acid–base equilibrium of an acid, HA, in water is shown below.

$$HA(aq) + H_2O(l) \rightleftharpoons H_3O^+(aq) + A^-(aq)$$

To emphasise the loss of H^+, this can be shown more simply as **dissociation** of the acid HA:

$$HA(aq) \rightleftharpoons H^+(aq) + A^-(aq).$$

The **strength** of an acid is the extent that the acid dissociates into H^+ and A^-.

A strong acid is completely dissociated.

Strong acids

A **strong** acid, such as nitric acid, HNO_3, is a **good** proton donor.
- There is almost **complete** dissociation.

$$HNO_3(aq) \xrightarrow{\text{equilibrium}} H^+(aq) + NO_3^-(aq)$$

- Virtually all of the potential acidic power has been released as $H^+(aq)$.
- At equilibrium, $[H^+(aq)]$ is much greater than $[HNO_3(aq)]$.

A weak acid only partially dissociates.

Weak acids

A **weak** acid, such as ethanoic acid, CH_3COOH, is a **poor** proton donor.
- There is only **partial** dissociation.

$$CH_3COOH(aq) \xleftarrow{\text{equilibrium}} H^+(aq) + CH_3COO^-(aq)$$

Only a small proportion of the potential acidic power has been released as $H^+(aq)$. At equilibrium, $[CH_3COOH(aq)]$ is much greater than $[H^+(aq)]$.

K_a is just a special equilibrium constant K_c for equilibria showing the dissociation of acids.

$[HA(aq)]$, $[H^+(aq)]$ and $[A^-(aq)]$ are equilibrium concentrations.

K_a is sometimes called the acidity constant.

The acid dissociation constant, K_a

The extent of acid dissociation is shown by the **acid dissociation constant**, K_a.

For the reaction: $HA(aq) \rightleftharpoons H^+(aq) + A^-(aq)$,

$$K_a = \frac{[H^+(aq)]\,[A^-(aq)]}{[HA(aq)]}$$

units: $K_a = \dfrac{(\text{mol dm}^{-3})^2}{(\text{mol dm}^{-3})} = \text{mol dm}^{-3}$.

- A **large** K_a value shows that the extent of **dissociation is large** – the acid is strong.
- A **small** K_a value shows that the extent of **dissociation is small** – the acid is weak.

Concentrated and *dilute* are terms used to describe the **amount** of dissolved acid in a solution.

Strong and *weak* are terms used to describe the degree of **dissociation** of an acid.

Acid strength and concentration

The distinction between the strength and concentration of an acid is important.

KEY POINT

Concentration is the **amount** of an acid dissolved in 1 dm^3 of solution.
- Concentration is measured in mol dm^{-3}.

Strength is the extent of **dissociation** of an acid.
- Strength is measured as K_a in units determined from the equilibrium.

Progress check

1 For each of the following acid–base equilibria, write down the expression for K_a. State the units of K_a for each reaction.
(a) $HCOOH(aq) \rightleftharpoons H^+(aq) + HCOO^-(aq)$
(b) $C_6H_5COOH(aq) \rightleftharpoons H^+(aq) + C_6H_5COO^-(aq)$

2 Samples of two acids, hydrochloric acid and ethanoic acid, have the same concentration: 0.1 mol dm^{-3}. Explain why one 'dilute acid' is *strong* whereas the other 'dilute acid' is *weak*.

1 (a) $K_a = \dfrac{[H^+(aq)]\,[HCOO^-(aq)]}{[HCOOH(aq)]}$ units: mol dm^{-3}
(b) $K_a = \dfrac{[H^+(aq)]\,[C_6H_5COO^-(aq)]}{[C_6H_5COOH(aq)]}$ units: mol dm^{-3}

2 Concentration applies to the amount, in mol, in 1 dm^3 of solution. Both solutions have 0.1 mol dissolved in 1 dm^3 of solution and are dilute. Hydrochloric acid is strong because its dissociation is near to complete. However, ethanoic acid only donates a small proportion of its potential protons, its dissociation is incomplete and it is a weak acid.

1.5 The pH scale

After studying this section you should be able to:

- *define the terms pH, pK_a and K_w*
- *calculate pH from $[H^+(aq)]$*
- *calculate $[H^+(aq)]$ from pH*
- *understand the meaning of the ionisation product of water, K_w*
- *calculate pH for strong bases*

LEARNING SUMMARY

pH and $[H^+(aq)]$

OCR MS
SALTERS MS

The concentrations of $H^+(aq)$ ions in aqueous solutions vary widely between about 10 mol dm^{-3} and about 1×10^{-15} mol dm^{-3}.

The **pH scale** is a logarithmic scale used to overcome the problem of using this large range of numbers and to ease the use of negative powers.

The pH scale	
pH	$[H^+]$ / mol dm^{-3}
0	1
1	1×10^{-1}
2	1×10^{-2}
3	1×10^{-3}
4	1×10^{-4}
5	1×10^{-5}
6	1×10^{-6}
7	1×10^{-7}
8	1×10^{-8}
9	1×10^{-9}
10	1×10^{-10}
11	1×10^{-11}
12	1×10^{-12}
13	1×10^{-13}
14	1×10^{-14}

KEY POINT

pH is defined as: $pH = -\log_{10} [H^+(aq)]$
- where $[H^+(aq)]$ is the concentration of hydrogen ions in aqueous solution.
- $[H^+(aq)]$ can be calculated from pH using:
  $[H^+(aq)] = 10^{-pH}$

Notice how the value of pH is linked to the power of 10.

pH	2	9	3.6	10.3
$[H^+(aq)]$/mol dm^{-3}	10^{-2}	10^{-9}	$10^{-3.6}$	$10^{-10.3}$

What does a pH value mean?

- A **low** value of $[H^+(aq)]$ matches a **high** value of pH.
- A **high** value of $[H^+(aq)]$ matches a **low** value of pH.
- A change of pH by 1 changes $[H^+(aq)]$ by 10 times.
- An acid of pH 4 contains 10 times the concentration of $H^+(aq)$ ions as an acid of pH 5.

Calculating the pH of strong acids

OCR MS
SALTERS MS

For a strong acid, HA:

- we can assume complete dissociation
- the concentration of $H^+(aq)$ can be found directly from the acid concentration: $[H^+] = [HA]$.

HA is **monoprotic**: each HA molecule can donate **one** H^+ ion.

You should be able to convert pH into $[H^+(aq)]$ and *vice versa.*

10^x

log

This is the key to use on your calculator when doing pH and $[H^+]$ calculations.

For 10^X, press the **SHIFT** or **INV** key first.

Example 1

A strong acid, HA, has a concentration of 0.010 mol dm^{-3}. What is the pH?

Complete dissociation. $\therefore [H^+(aq)] = 0.010$ mol dm^{-3}

$$pH = -\log_{10} [H^+(aq)] = -\log_{10} (0.010) = \mathbf{2.0}$$

Example 2

A strong acid, HA, has a pH of 3.4. What is the concentration of $H^+(aq)$?

Complete dissociation. $\therefore [H^+(aq)] = 10^{-pH} = 10^{-3.4}$ mol dm^{-3}

$$\therefore [H^+(aq)] = \mathbf{3.98 \times 10^{-4}} \text{ mol dm}^{-3}$$

Hints for pH calculations

Calculations involving pH are easy once you have learnt how to use your calculator properly.

- Try the examples on the previous page until you can remember the order to press the keys.
- Try reversing each calculation to go back to the original value. Repeat several times until you have mastered how to use **your** calculator for pH calculations.
- **Don't** borrow a calculator or you will get confused. Different calculators may need the keys to be pressed in a different order!
- Look at your answer and decide whether it looks sensible.

KEY POINT

Learn: $pH = -\log_{10} [H^+(aq)]$
$[H^+(aq)] = 10^{-pH}$.

Progress check

1 Calculate the pH of solutions with the following $[H^+(aq)]$ values.
(a) 0.01 mol dm^{-3}
(b) 0.0001 mol dm^{-3}
(c) 1.0×10^{-13} mol dm^{-3}
(d) 2.50×10^{-3} mol dm^{-3}
(e) 8.10×10^{-6} mol dm^{-3}
(f) 4.42×10^{-11} mol dm^{-3}

2 Calculate $[H^+(aq)]$ of solutions with the following pH values.
(a) pH 3 (b) pH 10 (c) pH 2.8 (d) pH 7.9 (e) pH 12.2 (f) pH 9.6

3 How many times more hydrogen ions are in an acid of pH 1 than an acid of pH 5?

4 How can a solution have a pH with a negative value?

1 (a) 2 (b) 4 (c) 13 (d) 2.60 (e) 5.09 (f) 10.4
2 (a) 1×10^{-3} mol dm^{-3} (b) 1×10^{-10} mol dm^{-3} (c) 1.58×10^{-3} mol dm^{-3} (d) 1.26×10^{-8} mol dm^{-3} (e) 6.31×10^{-13} mol dm^{-3} (f) 2.51×10^{-10} mol dm^{-3}
3 pH 1 has 10 000 times more H^+ ions than pH 5.
4 A solution with $[H^+(aq)] > 1$ mol dm^{-3} has a negative pH value.

Calculating the pH of weak acids

OCR MS
SALTERS MS

The pH of a weak acid HA can be calculated from:

- the **concentration** of the acid and
- the value of the acid dissociation constant, K_a.

Assumptions and approximations

Consider the equilibrium of a weak aqueous acid HA(aq):

$$HA(aq) \rightleftharpoons H^+(aq) + A^-(aq)$$

- Assuming that only a very small proportion of HA dissociates, the equilibrium concentration of HA(aq) will be very nearly the same as the concentration of undissociated HA(aq).

$$\therefore [HA(aq)]_{equilibrium} \approx [HA(aq)]_{start}$$

- Assuming that there is a negligible proportion of $H^+(aq)$ from ionisation of water:

$$[H^+(aq)] \approx [A^-(aq)]$$

- Using these approximations

$$K_a = \frac{[H^+(aq)]\,[A^-(aq)]}{[HA(aq)]} \qquad \therefore K_a \approx \frac{[H^+(aq)]^2}{[HA(aq)]}$$

For calculations, use $K_a \approx \dfrac{[H^+(aq)]^2}{[HA(aq)]}$

Take care to learn this method.

Many marks are dropped on exam papers by students who have not done so!

Example

For a weak acid [HA(aq)] = 0.100 mol dm^{-3}, K_a = 1.70 × 10^{-5} mol dm^{-3} at 25°C. Calculate the pH.

$$K_a = \frac{[H^+(aq)]\,[A^-(aq)]}{[HA(aq)]} \approx \frac{[H^+(aq)]^2}{[HA(aq)]}$$

$$\therefore\ 1.70 \times 10^{-5} = \frac{[H^+(aq)]^2}{0.100}$$

$$\therefore\ [H^+(aq)] = \sqrt{0.100 \times 1.70 \times 10^{-5}} = 0.00130 \text{ mol dm}^{-3}$$

$$pH = -\log_{10}[H^+(aq)] = -\log_{10}(0.00130) = \mathbf{2.89}$$

K_a and pK_a

OCR M5
SALTERS M5

Values of K_a can be made more manageable if expressed in a logarithmic form, pK_a (see also page 31: pH and [H$^+$(aq)]).

$$\mathbf{pK_a = -\log_{10}K_a}$$
$$\mathbf{K_a = 10^{-pKa}}$$

K_a and pK_a conversions are just like those between pH and H$^+$.

- A **low** value of K_a matches a **high** value of pK_a
- A **high** value of K_a matches a **low** value of pK_a

The smaller the pK_a value, the stronger the acid.

Comparison of K_a and pK_a

acid		*K_a / mol dm^{-3}*	*pK_a*
methanoic acid	HCOOH	1.6 × 10^{-4}	$-\log_{10}$(1.6 × 10^{-4}) = **3.8**
benzoic acid	C_6H_5COOH	6.3 × 10^{-5}	$-\log_{10}$(6.3 × 10^{-5}) = **4.2**

Progress check

1 Find the pH of solutions of a weak acid HA (K_a = 1.70 × 10^{-5} mol dm^{-3}), with the following concentrations:
(a) 1.00 mol dm^{-3}; (b) 0.250 mol dm^{-3}; (c) 3.50 × 10^{-2} mol dm^{-3}

2 Find values of K_a and pK_a for the following weak acids.
(a) 1.0 mol dm^{-3} solution with a pH of 4.5
(b) 0.1 mol dm^{-3} solution with a pH of 2.2
(c) 2.0 mol dm^{-3} solution with a pH of 3.8.

1 (a) 2.38 (b) 2.69 (c) 3.11
2 (a) K_a = 1 × 10^{-9} mol dm^{-3}, pK_a = 9
(b) K_a 3.98 × 10^{-4} mol dm^{-3}, pK_a = 3.4
(c) K_a 1.26 × 10^{-8} mol dm^{-3}, pK_a = 7.9

The ionisation of water and K_w

OCR M5
SALTERS M5

In water, a very small proportion of molecules dissociates into H$^+$(aq) and OH$^-$(aq) ions. The position of equilibrium lies well to the left of the equation below, representing this dissociation.

$$\xleftarrow{\text{equilibrium}}$$
$$H_2O(l) \rightleftharpoons H^+(aq) + OH^-(aq)$$

Treating water as a weak acid: $K_a = \dfrac{[H^+(aq)]\,[OH^-(aq)]}{[H_2O(l)]}$

Rearranging gives:

$$\underbrace{K_a \times [H_2O(l)]}_{\text{constant, } K_w} = [H^+(aq)]\,[OH^-(aq)]$$

[H_2O(l)] is constant and is included within K_w

KEY POINT

The constant K_w is called the **ionic product of water**
- K_w = [H$^+$(aq)] [OH$^-$(aq)]
- At 25°C, K_w = 1.0 × 10^{-14} mol^2 dm^{-6}.

In water, the concentrations of $H^+(aq)$ and $OH^-(aq)$ ions are the same.

- $[H^+(aq)] = [OH^-(aq)] = 10^{-7}$ mol dm^{-3} $(10^{-14} = 10^{-7} \times 10^{-7})$

Hydrogen ion and hydroxide ion concentrations

Linking $[H^+]$ and $[OH^-]$

$10^{-14} = [H^+][OH^-]$

Water: pH = 7,
$[H^+] = 10^{-7}$ mol dm^{-3}
$10^{-14} = 10^{-7} \times 10^{-7}$

An acid: pH = 3
$[H^+] = 10^{-3}$ mol dm^{-3}
$10^{-14} = 10^{-3} \times 10^{-11}$

An alkali: pH = 10
$[H^+] = 10^{-10}$ mol dm^{-3}
$10^{-14} = 10^{-10} \times 10^{-4}$

All aqueous solutions contain $H^+(aq)$ and $OH^-(aq)$ ions. The proportions of these ions in a solution are determined by the pH.

In water	$[H^+(aq)] = [OH^-(aq)]$
In **acidic** solutions	$[H^+(aq)] > [OH^-(aq)]$
In **alkaline** solutions	$[H^+(aq)] < [OH^-(aq)]$

The concentrations of $H^+(aq)$ and $OH^-(aq)$ are linked by K_w.

- At 25°C $1.0 \times 10^{-14} = [H^+(aq)][OH^-(aq)]$
- The indices of $[H^+(aq)]$ and $[OH^-(aq)]$ add up to –14.

Calculating the pH of strong alkalis

The pH of a strong alkali can be found using K_w.

For a strong alkali, BOH:

- we can assume complete dissociation
- the concentration of $OH^-(aq)$ can be found directly from the alkali concentration: $[BOH] = [OH^-]$.

To find the pH of an alkali, first find $[H^+]$ using K_w and $[OH^-]$.

Example

A strong alkali, BOH, has a concentration of 0.50 mol dm^{-3}.

What is the pH?

Complete dissociation. $\therefore [OH^-(aq)] = 0.50$ mol dm^{-3}

$$K_w = [H^+(aq)][OH^-(aq)] = 1 \times 10^{-14} \text{ mol}^2 \text{ dm}^{-6}$$

$$\therefore [H^+(aq)] = \frac{K_w}{[OH^-(aq)]} = \frac{1 \times 10^{-14}}{0.50} = 2 \times 10^{-14} \text{ mol dm}^{-3}$$

$$pH = -\log_{10}[H^+(aq)] = -\log_{10}(2 \times 10^{-14}) = \mathbf{13.7}$$

Progress check

1 Find the $[H^+(aq)]$ and pH of the following alkalis at 25°C:
(a) 1×10^{-3} mol dm^{-3} $OH^-(aq)$
(b) 3.5×10^{-2} mol dm^{-3} $OH^-(aq)$.

2 Find the pH of the following solutions of strong bases at 25°C:
(a) 0.01 mol dm^{-3} KOH (aq)
(b) 0.20 mol dm^{-3} NaOH (aq).

1 (a) $[H^+(aq)] = 1 \times 10^{-11}$ mol dm^{-3}; pH = 11
(b) $[H^+(aq)] = 2.86 \times 10^{-13}$ mol dm^{-3}; pH = 12.5.

2 (a) pH = 12; (b) pH = 13.3.

1.6 pH changes

LEARNING SUMMARY

After studying this section you should be able to:

- *recognise the shapes of titration curves for acids and bases with different strengths*
- *explain the choice of suitable indicators for acid–base titrations*
- *carry out unstructured titration calculations*
- *state what is meant by a buffer solution*
- *explain how pH is controlled by each component in a buffer solution*
- *calculate the pH of a buffer solution*

Titration curves

OCR M5

Choosing an indicator using titration curves

A titration curve shows the changes in pH during a titration.

In the titration curves below, different combinations of strong and weak acids and alkalis have been used.

- The end point of a titration is identified by the colour change of the indicator.
- Different indicators change colour at different pH values.
- The pH values at which the indicators methyl orange (**MO**) and phenolphthalein (**P**) change colour are shown on each diagram.

strong acid/strong alkali

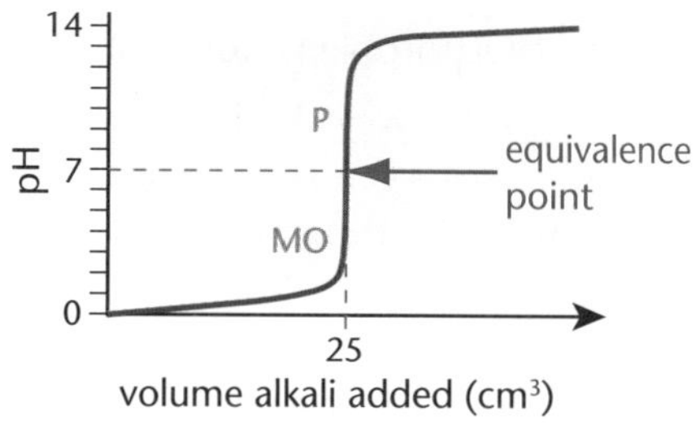

both indicators suitable

strong acid/weak alkali

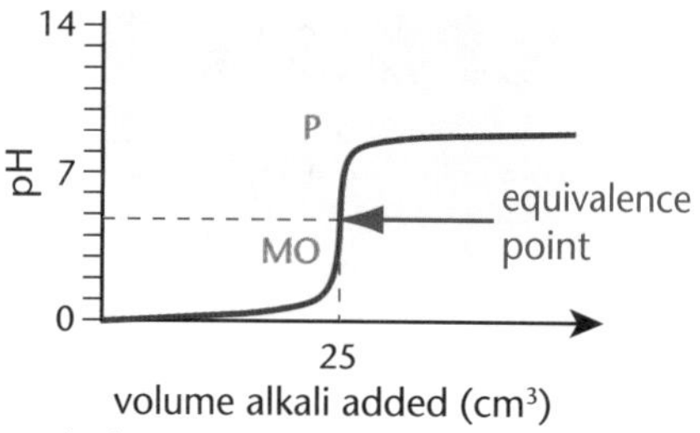

methyl orange (**MO**) *suitable*
phenolphthalein (**P**) *unsuitable*

weak acid/strong alkali

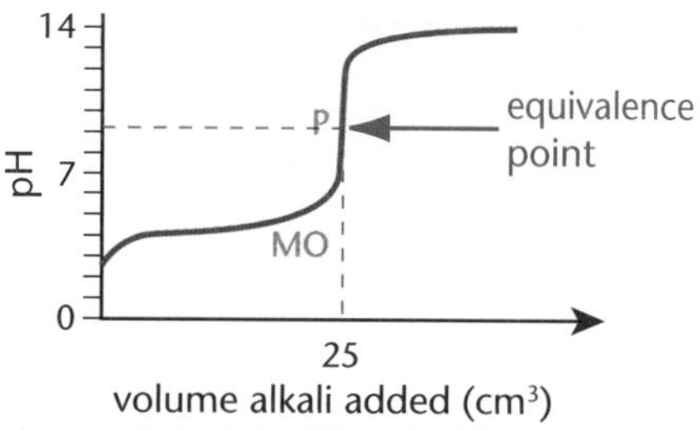

phenolphthalein (**P**) *suitable*
methyl orange (**MO**) *unsuitable*

weak acid/weak alkali

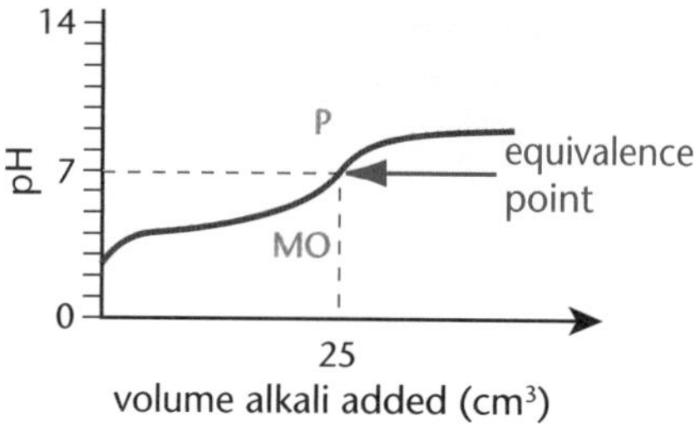

neither indicator suitable

Choose the correct indicator

Strong acid/strong alkali
phenolphthalein ✓
methyl orange ✓

Strong acid/weak alkali
phenolphthalein ✗
methyl orange ✓

Weak acid/strong alkali
phenolphthalein ✓
methyl orange ✗

Weak acid/weak alkali
phenolphthalein ✗
methyl orange ✗

Key features of titration curves

- The pH changes rapidly at the near vertical portion of the titration curve. The mid-point of the vertical portion is the **equivalence point** of the titration.
- The sharp change in pH is brought about by a very small addition of alkali, typically the addition of one drop.
- The indicator is only suitable if its pK_{ind} value is within the pH range of the near vertical portion of the titration curve.

Enthalpy changes during neutralisation

OCR M5

The *standard enthalpy change of neutralisation* $\Delta H^{\ominus}_{neut}$ is the energy change that accompanies the neutralisation of an acid by a base to form 1 mole of $H_2O(l)$, under standard conditions.

HCl(aq)	+	NaOH(aq)	$\rightleftharpoons$	NaCl(aq)	+	$H_2O(l)$ $\Delta H^{\ominus}_{neut} = -57.9 \text{ kJ mol}^{-1}$
acid		base				1 mol

The ionic equation for this reaction is simply:

$$H^+(aq) + OH^-(aq) \longrightarrow H_2O(l).$$

- In aqueous solution, strong acids and bases are completely dissociated and $\Delta H^{\ominus}_{neut}$ is approximately equal to the value above.
- For weak acids, this enthalpy change is less exothermic because some input of energy is required to dissociate the acid (see also pages 29–30).

Unstructured titration calculations

OCR M5
SALTERS M4

AS exam questions would help you through titration calculations by asking each step as a question.

But at A2 you are really on your own! You should now know more chemistry than at AS and also be able to tackle chemistry problems independently.

At least that is what the setters of examination questions expect!

In AS chemistry examinations, titration problems are usually highly structured. You are helped through calculations with each step being asked in a sequence that leads to the final solution.

In A2 chemistry examinations:

- problems will be set in which you are only provided with a summary of the experimental results
- you will be expected to navigate your own way through the problem; you will have no hints about how to solve the problem.

The key to success is a good understanding of the stages required to solve a titration problem. Most problems follow a common framework.

Acid–base titrations are always likely to be carried out in chemical analysis. In A2, you will also come across redox titrations (see pages 79–81). Although these are different types of titration, the principles for solving any calculations are the same.

Stage 1

All titrations use a standard solution – one whose concentration is known. In the titration, you measure the volume of this solution that exactly reacts with a second solution.

Remember that the steps shown here are key to solving most titration calculations.

See also redox titrations, pages 79–81.

- You first find the amount, in moles, of the compound in the standard solution.

E.g. In a titration 25.0 cm^3 of 0.100 mol dm^{-3} NaOH reacted exactly with a solution of H_2SO_4.

$$n = c \times \frac{V}{1000} = 0.100 \times \frac{25.0}{1000} = \mathbf{0.00250\ mol}$$

Stage 2

- The equation is then usually used to find the number of moles of the other substance dissolved in the second solution:

E.g.	$2NaOH(aq)$ +	$H_2SO_4(aq) \longrightarrow$ *products*
	2 mol	**1 mol**
amounts:	**0.00250**	**0.00125 mol**

Stage 3

- Finally, this value is processed in some way to find out some further information about the second substance. This is the hardest stage and it may contain several steps. This stage may also vary between different problems.

Worked example

Compound **A** is a straight-chain carboxylic acid. A student analysed a sample of acid **A** by the procedure below.

The student dissolved 5.437 g of **A** in water and made the solution up to 250 cm^3.

In a titration, 25.0 cm^3 of 0.200 mol dm^{-3} NaOH were neutralised by exactly 23.45 cm^3 of solution **A**.

Use the results to calculate the molar mass of acid **A** and suggest its identity.

The concentration and volume of NaOH are known. So we work out the amount of NaOH first.

The amount of NaOH used can be calculated from the titration results:

amount of NaOH = $0.200 \times \frac{25.0}{1000}$ = **0.00500 mol**

The amount of carboxylic acid that reacts with the NaOH can be determined from the balanced equation for the reaction:

Carboxylic acid **A** must have the formula: RCOOH

The equation and molar reacting quantities are:

In this titration there is the same number of moles of each reactant. We work this out from the balanced equation.

$$NaOH(aq) + RCOOH(aq) \longrightarrow RCOONa(aq) + H_2O(l)$$

1 mol **1 mol**

∴ in the titration: 0.00500 mol NaOH reacts with **0.00500 mol RCOOH**

You must now look at what the question is asking. You need to think about how you are going to solve the rest of the problem.

- Here, we must first find out the amount, in moles, of **A** that was used to prepare the original 250 cm^3 solution.
- In the titration, 23.45 cm^3 of this solution of **A** was used – we know the number of moles of **A** dissolved in this volume.
- We need to scale up these quantities to find the number of moles of **A** dissolved in 250 cm^3 of solution:

This step is important but many candidates forget to do this (or don't understand that it must be done!)

23.45 cm^3 of RCOOH(aq) in the titration contains 0.00500 mol RCOOH

1 cm^3 of RCOOH(aq) in the titration contains $\frac{0.00500}{23.45}$ mol RCOOH

250 cm^3 of RCOOH(aq) contains $\frac{0.00500}{23.45} \times 250$ = 0.0533 mol RCOOH

This step is easy, provided that you have remembered the following:
$n = \frac{\text{mass } m}{\text{Molar mass } M}$

The molar mass of RCOOH can now be determined and we have solved the first part of the problem:

$n = \frac{m}{M}$ ∴ Molar mass, $M = \frac{m}{n} = \frac{5.437}{0.0533}$ = **102.0 g mol^{-1}**

*Finally, we need to suggest a possible structure for straight-chain carboxylic acid **A**.*

The formula of the straight-chain carboxylic acid is RCOOH.

These problems may mix together chemistry from different parts of your synoptic assessment. See pages 160–163.

COOH has a relative mass of 12.0 + 16.0 x 2 + 1.0 = 45.0.

∴ the alkyl group R has a relative mass of 102.0 – 45.0 = 57.0

∴ the alkyl group must be $CH_3CH_2CH_2CH_2$ (15.0 + 14.0 + 14.0 + 14.0)

and carboxylic acid **A** is **$CH_3CH_2CH_2CH_2COOH$** (pentanoic acid).

Buffer solutions

OCR M5
SALTERS M5

A buffer solution minimises changes in pH during the addition of an acid or an alkali to a solution. The buffer solution maintains a near-constant pH by removing most of any added acid or alkali.

A **buffer solution** is a mixture of:

- a **weak acid**, HA, and
- its **conjugate base**, A^-:

$$\underset{\text{weak acid}}{HA(aq)} \rightleftharpoons H^+(aq) + \underset{\text{conjugate base}}{A^-(aq)}$$

Buffers are added to foods to prevent deterioration due to pH change (caused by bacterial or fungal activity).

In a buffer solution, the concentration of hydrogen ions, $[H^+(aq)]$, is very small compared with the concentrations of the weak acid $[HA(aq)]$ or the conjugate base $[A^-(aq)]$.

$$[H^+(aq)] \ll [HA(aq)] \quad \text{and} \quad [H^+(aq)] \ll [A^-(aq)].$$

How does a buffer act?

We can explain how a buffer minimises pH changes by using le Chatelier's principle.

Addition of an acid, $H^+(aq)$, to a buffer

On addition of an acid:

- $[H^+(aq)]$ is increased
- the pH change is opposed and the **equilibrium moves to the left, removing $[H^+(aq)]$** and forming HA(aq)
- the **conjugate base $A^-(aq)$** removes **most** of any added $[H^+(aq)]$.

A^- removes most of any added acid.

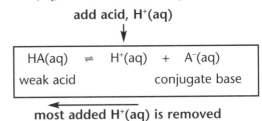

add acid, $H^+(aq)$

$$\underset{\text{weak acid}}{HA(aq)} \rightleftharpoons H^+(aq) + \underset{\text{conjugate base}}{A^-(aq)}$$

← **most added $H^+(aq)$ is removed**

Addition of an alkali, $OH^-(aq)$, to a buffer

On addition of an alkali:

- the added $OH^-(aq)$ reacts with the small concentration of $H^+(aq)$:

$$H^+(aq) + OH^-(aq) \longrightarrow H_2O(l)$$

- the pH change is opposed – the **equilibrium moves to the right, restoring $[H^+(aq)]$** as HA(aq) dissociates
- the **weak acid HA** restores **most** of any $[H^+(aq)]$ that has been removed.

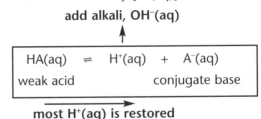

add alkali, $OH^-(aq)$

$$\underset{\text{weak acid}}{HA(aq)} \rightleftharpoons H^+(aq) + \underset{\text{conjugate base}}{A^-(aq)}$$

→ **most $H^+(aq)$ is restored**

HA removes added alkali.

KEY POINT

Although the two components in a buffer solution react with added acid and alkali, they **cannot stop** the pH from changing. They do however **minimise** pH changes.

Remember these buffers:
CH_3COOH/CH_3COONa
and
NH_4Cl/NH_3

The pH of healthy blood is maintained at between 7.35 and 7.40 by the carbonic acid–hydrogencarbonate buffer system.
H_2CO_3 is the **weak acid**,
HCO_3^- is the **conjugate base**.
$H_2CO_3(aq) \rightleftharpoons H^+(aq) + HCO_3^-(aq)$
Blood cannot sustain life if its pH lies much outside of this narrow range.

Common buffer solutions

An acidic buffer

A common acidic buffer is an aqueous solution containing a mixture of ethanoic acid, CH_3COOH, and the ethanoate ion, CH_3COO^-.

- Ethanoic acid acts as the **weak acid**, CH_3COOH.
- Sodium ethanoate, $CH_3COO^-Na^+$, acts as a source of the **conjugate base**, CH_3COO^-.

An alkaline buffer

A common alkaline buffer is an aqueous solution containing a mixture of the ammonium ion, NH_4^+, and ammonia, NH_3.

- Ammonium chloride, $NH_4^+Cl^-$, acts as source of the **weak acid**, NH_4^+.
- Ammonia acts as the **conjugate base**, NH_3.

Calculations involving buffer solutions

OCR MS
SALTERS MS

For buffer calculations, learn:

$$[H^+(aq)] = K_a \times \frac{[HA(aq)]}{[A^-(aq)]}$$

The pH of a buffer can be controlled by the acid/base ratio.

The pH of a buffer solution depends upon:

- the acid dissociation constant, K_a, of the buffer system
- the **ratio** of the weak **acid** and its conjugate **base**.

For a buffer comprising the weak acid, HA, and its conjugate base, A^-,

$$K_a = \frac{[H^+(aq)]\,[A^-(aq)]}{[HA(aq)]}$$

$$\therefore\ [H^+(aq)] = K_a \times \frac{[HA(aq)]}{[A^-(aq)]}$$

K_a ← acid dissociation constant

$\frac{[HA(aq)]}{[A^-(aq)]}$ ← **ratio** of the weak acid and its conjugate base

Example

Calculate the pH of a buffer comprising 0.30 mol dm^{-3} $CH_3COOH(aq)$ ($K_a = 1.7 \times 10^{-5}$ mol dm^{-3}) and 0.10 mol dm^{-3} $CH_3COO^-(aq)$.

$$[H^+(aq)] = K_a \times \frac{[HA(aq)]}{[A^-(aq)]}$$

$$\therefore\ [H^+(aq)] = 1.7 \times 10^{-5} \times \frac{0.30}{0.10} = 5.1 \times 10^{-5}\ \text{mol dm}^{-3}$$

$$\therefore\ pH = -\log_{10}[H^+(aq)] = -\log_{10}(5.1 \times 10^{-5})$$

$\therefore$ pH of the buffer solution = **4.3**

Progress check

1 Three buffer solutions are made from benzoic acid, C_6H_5COOH and sodium benzoate, C_6H_5COONa with the following compositions:

Buffer A: 0.10 mol dm^{-3} C_6H_5COOH and 0.10 mol dm^{-3} C_6H_5COONa
Buffer B: 0.75 mol dm^{-3} C_6H_5COOH and 0.25 mol dm^{-3} C_6H_5COONa
Buffer C: 0.20 mol dm^{-3} C_6H_5COOH and 0.80 mol dm^{-3} C_6H_5COONa

(a) Write the equation for the equilibrium in these buffers.
(b) Write an expression for K_a of benzoic acid.
(c) Calculate the pH of each of the buffer solutions (K_a for $C_6H_5COOH = 6.3 \times 10^{-5}$ mol dm^{-3}.)

1 (a) $C_6H_5COOH \rightleftharpoons H^+(aq) + C_6H_5COO^-(aq)$
(b) $K_a = \frac{[H^+(aq)]\,[C_6H_5COO^-(aq)]}{[C_6H_5COOH(aq)]}$
(c) Buffer **A**: 4.2 Buffer **B**: 3.7 Buffer **C**: 4.8

Sample question and model answer

Many cosmetics contain buffers. Human skin is slightly acidic and has a pH of approximately 5.50. Skin-care products often buffer at a pH 5.50 or below.

(a) What do you understand by the term *buffer*?

A solution that minimises a change in pH. ✓ [1]

Don't use the phrases 'keeps the pH constant' or 'stops the pH from changing'.
The pH will change but the buffer keeps the change small.

(b) Lactic acid, $CH_3CH(OH)COOH$, is used in many cosmetics. Lactic acid is a weak acid with an acid dissociation constant, K_a, of 8.40×10^{-4} mol dm^{-3}.

(i) Write an equation to show the dissociation of lactic acid into its ions.

$CH_3CH(OH)COOH \rightleftharpoons CH_3CH(OH)COO^- + H^+$ ✓ (equilibrium sign essential)

(ii) Write an expression for the acid dissociation constant, K_a, of lactic acid.

$$K_a = \frac{[CH_3CH(OH)COO^-(aq)]\ [H^+(aq)]}{[CH_3CH(OH)COOH(aq)]}$$ ✓

(iii) Calculate the pH of a 1.25×10^{-2} mol dm^{-3} solution of lactic acid.

$CH_3CH(OH)COO^- \approx H^+$.

equilibrium conc. of weak acid ≈ undissociated concentration of weak acid.

These approximations simplify the K_a expression to:
$K_a \approx \frac{[H^+]^2}{[HA]}$

$$\therefore K_a \approx \frac{[H^+(aq)]^2}{[CH_3CH(OH)COOH(aq)]} \qquad 8.4 \times 10^{-4} = \frac{[H^+(aq)]^2}{1.25 \times 10^{-2}}$$ ✓

$[H^+(aq)] = \sqrt{(1.25 \times 10^{-2} \times 8.40 \times 10^{-4})} = 3.24 \times 10^{-3}$ ✓ mol dm^{-3}

$\therefore$ pH $= -\log(3.24 \times 10^{-3}) = 2.49$. ✓ [5]

Make sure that your calculator skills are good and show your working. A correct method will secure most available marks.
Notice the use throughout of 3 significant figures (in the question and in the answers).

(c) A buffer solution can be made based on lactic acid and a salt of lactic acid. Explain how this buffer solution acts as a buffer.

In the equilibrium $CH_3CH(OH)COOH \rightleftharpoons CH_3CH(OH)COO^- + H^+$, there are large excesses of $CH_3CH(OH)COOH$ and $CH_3CH(OH)COO^-$. ✓

The equilibrium above shifts in response to added acid and alkali. ✓

On addition of an alkali, OH^- ions react with the small concentration of H^+ present:

$H^+(aq) + OH^-(aq) \longrightarrow H_2O(l)$

The equilibrium shifts to the right, restoring most of any H^+ ions removed: ✓

$CH_3CH(OH)COOH \longrightarrow CH_3CH(OH)COO^- + H^+$ ✓

Lactate ions from sodium lactate remove most of any added acid, shifting the equilibrium to the left: ✓

$CH_3CH(OH)COO^- + H^+ \longrightarrow CH_3CH(OH)COOH$ ✓ [6]

Practise explaining how a buffer works. This is often asked in exams and it is harder to answer than it appears.
You need to discuss shifts in the equilibrium between the weak acid and its conjugate base.

(d) A buffer solution based on lactic acid is required to buffer at a pH 3.80. Calculate the ratio of lactic acid : sodium lactate that will be needed to give this pH.

In the buffer, $[H^+(aq)] = 10^{-3.80} = 1.58 \times 10^{-4}$ ✓ mol dm^{-3}

In the buffer, $[H^+(aq)] = K_a \times \frac{[CH_3CH(OH)COOH(aq)]}{[CH_3CH(OH)COO^-(aq)]}$ ✓

$$\therefore \text{ratio } \frac{[CH_3CH(OH)COOH(aq)]}{[CH_3CH(OH)COO^-(aq)]} = \frac{[H^+(aq)]}{K_a} = \frac{1.58 \times 10^{-4}}{8.4 \times 10^{-4}}$$ ✓

ratio lactic acid : sodium lactate $= 0.189 = \frac{1}{5.30}$ ✓ [4]

[Total: 16]

The calculated ratio is 0.189.
To get $\frac{1}{5.30}$ you need to press the x^{-1} key.
The answer means that you need 5.3 times the concentration of sodium lactate to lactic acid.

Practice examination questions

1 The reaction between hydrogen peroxide and iodide ions in an acid solution can be written as follows.

$$H_2O_2(aq) + 2I^-(aq) + 2H^+(aq) \longrightarrow I_2(aq) + 2H_2O(l)$$

(a) The table below shows initial rates from different initial concentrations of H_2O_2(aq), I^-(aq) and H^+(aq) at constant temperature.

[H_2O_2(aq)] /mol dm^{-3}	*[I^-(aq)] /mol dm^{-3}*	*[H^+(aq)] /mol dm^{-3}*	*initial rate /10^{-6} mol dm^{-3} s^{-1}*
0.00075	0.10	0.10	2.1
0.00150	0.10	0.10	4.2
0.00150	0.10	0.20	4.2
0.00075	0.70	0.10	14.7

(i) Determine, with reasoning, the order with respect to each reactant.

(ii) Write the rate equation and calculate the value of the rate constant, *k*, for this reaction. [9]

(b) (i) What is meant by the term *rate-determining step*?

(ii) Using the rate equation in (a)(ii), suggest an equation for the rate-determining step of this reaction. [2]

[Total: 11]

2 (a) A chemical reaction is second order with respect to compound **A** and first order with respect to compound **B**.

(i) Write the rate equation for this reaction.

(ii) What is the overall order of this reaction?

(iii) By what factor will the rate increase if the concentrations of **A** and **B** are both tripled?

(iv) When [**A**] = 0.40 mol dm^{-3} and **B** = 0.35 mol dm^{-3}, the rate = 1.76 x 10^{-4} mol dm^{-3} s^{-1}. Calculate the value of the rate constant, stating its units. [5]

(b) Propanone and iodine react together in acidic solution to give iodopropanone according to the overall equation below.

$$CH_3COCH_3(aq) + I_2(aq) \longrightarrow CH_3COCH_2I(aq) + HI(aq)$$

The rate equation is: *rate* = *k*[CH_3COCH_3(aq)] [H^+(aq)]

(i) State, with an explanation, the role of the H^+(aq) ions in this reaction.

(ii) Suggest a possible equation for the rate-determining step. [3]

[Total: 8]

3 The equilibrium below was set up.

$$CH_3COOCH_2CH_3 + H_2O \rightleftharpoons CH_3COOH + CH_3CH_2OH$$

The equilibrium mixture contains 1.35 mol of ethyl ethanoate, 1.35 mol of water, 0.15 mol of ethanoic acid, and 3.15 mol of ethanol. It has a volume of 353 cm^3.

(a) Calculate K_c for this equilibrium to 2 significant figures. State the units, if any. [3]

(b) More ethanol was to be added to the equilibrium mixture.

(i) What would happen to the equilibrium?
Explain your reasoning.

(ii) What would be the effect of this change on the value of K_c? Explain your reasoning. [4]

[Total: 7]

Practice examination questions (continued)

4 When 0.60 moles of $H_2(g)$ and 0.18 moles of $I_2(g)$ were heated to constant temperature in a sealed container with a volume of 1 dm^3, an equilibrium was set up:

$$H_2(g) + I_2(g) \rightleftharpoons 2HI(g)$$

At equilibrium, 0.16 mol $I_2(g)$ had reacted.

(a) (i) Determine the equilibrium concentrations of H_2, I_2, and HI.

(ii) Calculate the equilibrium constant, K_c, for this reaction. [6]

(b) When the equilibrium mixture was heated by 100°C the value of K_c decreased. What additional information is provided by this observation? [2]

[Total: 8]

5 Nitric acid, HNO_3 is a strong acid and methanoic acid, HCOOH, is a weak acid ($K_a = 1.6 \times 10^{-4}$ mol dm^{-3}).

(a) Define the term *Brønsted–Lowry acid.* [1]

(b) Write the acid–base equilibrium that would be set up between methanoic acid and nitric acid. State the role of methanoic acid in this equilibrium. [2]

(c) Calculate the pH of the following solutions at 25°C:

(i) 0.176 mol dm^{-3} nitric acid

(ii) 0.372 mol dm^{-3} potassium hydroxide

(iii) 0.263 mol dm^{-3} methanoic acid.

(iv) a solution containing 0.255 mol dm^{-3} methanoic acid and 0.325 sodium methanoate. [10]

[Total: 13]

6 (a) (i) What is meant by the terms *strong acid* and *weak acid*?

(ii) Define the term *Brønsted–Lowry base.* [2]

(b) 25.0 cm^3 of a solution of sodium hydroxide contained 1.50×10^{-3} mol of NaOH.

(i) Calculate the pH of this solution at 298 K.

(ii) A student added 50.0 cm^3 of 0.250 mol dm^{-3} HCl to the 25.0 cm^3 solution of sodium hydroxide.

Calculate the pH of the solution formed. [8]

(c) A sample of hydrochloric acid had a pH of 1.52. A 25 cm^3 sample of this acid was diluted with 15 cm^3 of water.

Calculate the pH of the resulting solution. [4]

[Total: 14]

Chapter 2

Energy changes in chemistry

The following topics are covered in this chapter:

- Enthalpy changes
- Entropy changes
- Electrochemical cells
- Predicting redox reactions

2.1 Enthalpy changes

After studying this section you should be able to:

LEARNING SUMMARY

- explain and use the term 'lattice enthalpy'
- construct Born–Haber cycles to calculate the lattice enthalpy of simple ionic compounds
- explain the effect of ionic charge and ionic radius on the numerical magnitude of a lattice enthalpy
- calculate enthalpy changes of solution for ionic compounds from enthalpy changes of hydration and lattice enthalpies

Key points from AS

- **Enthalpy changes**
Revise AS pages 74–82

During the study of enthalpy changes in AS Chemistry, you learnt how to:

- calculate enthalpy changes directly from experiments using the relationship $Q = mc\Delta T$
- calculate enthalpy changes indirectly using Hess' law.

These key principles are built upon by considering the enthalpy changes that bond together an ionic lattice.

Lattice enthalpy

OCR M5
SALTERS M5

Ionic bonding is the electrostatic attraction between oppositely charged ions. Lattice enthalpy indicates the strength of the ionic bonds in an ionic lattice.

Key points from AS

- **Ionic bonding**
Revise AS pages 45–49

KEY POINT

The lattice enthalpy ($\Delta H^{\ominus}_{L.E.}$) of an ionic compound is the enthalpy change that accompanies the formation of 1 mole of an ionic compound from its constituent gaseous ions. ($\Delta H^{\ominus}_{L.E.}$ is **exothermic**.)

$$Na^+(g) + Cl^-(g) \longrightarrow Na^+Cl^-(s)$$

Each ion is surrounded by oppositely-charged ions, forming a giant ionic lattice.

Lattice enthalpy is also sometimes called the **enthalpy change of lattice formation.**

The opposite change, to break up the lattice, is called the **enthalpy change of lattice dissociation.**

$$Na^+Cl^-(s) \longrightarrow Na^+(g) + Cl^-(g)$$

Determination of lattice enthalpies

OCR M5

Lattice enthalpies cannot be determined directly and must be calculated indirectly using Hess' Law from other enthalpy changes that can be found experimentally. The energy cycle used to calculate a lattice enthalpy is the **Born–Haber cycle.**

The basis of the Born–Haber cycle is the formation of an ionic lattice from its element by two routes.

The Born–Haber cycle for sodium chloride is shown below.

Definitions for these enthalpy changes are shown below.

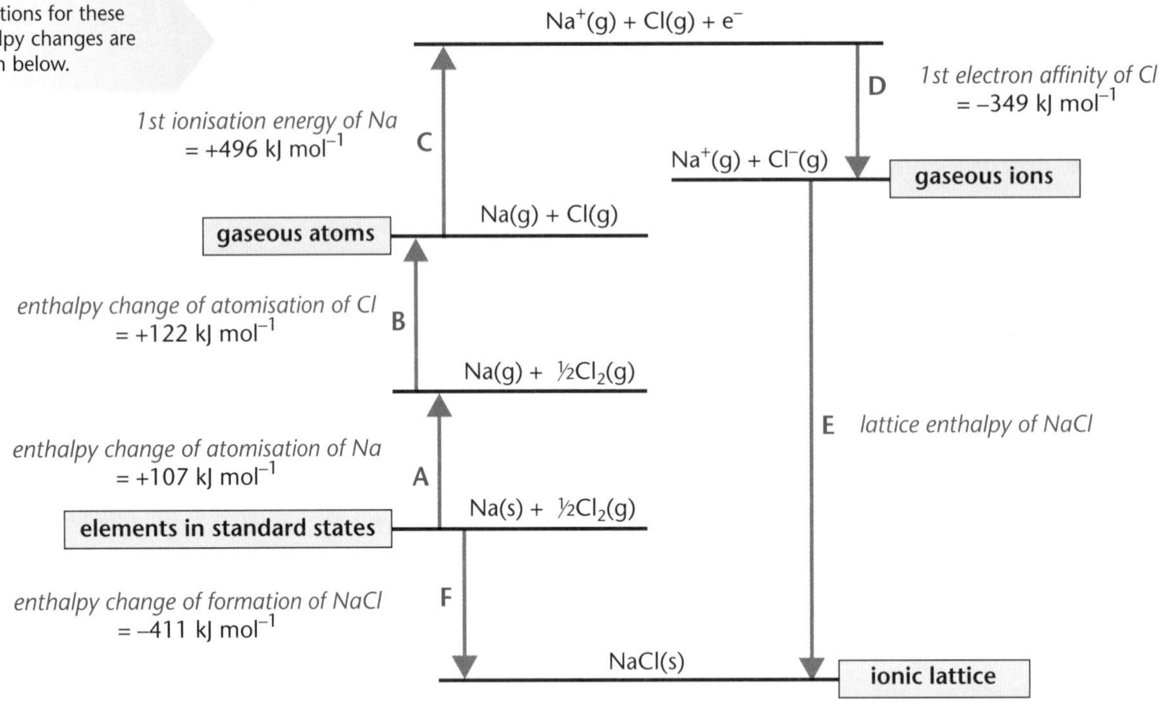

- the 1st ionisation energy of Na is **endothermic**
- the 1st electron affinity of Cl is **exothermic**.

Route 1: **A + B + C + D + E** *Route 2:* **F**

Using Hess' Law: **A + B + C + D + E = F**

$\Delta H^\ominus_{at}$ Na(g) + $\Delta H^\ominus_{at}$ Cl(g) + $\Delta H^\ominus_{I.E.}$Na(g) + $\Delta H^\ominus_{E.A.}$Cl(g) + **E** = $\Delta H^\ominus_f$ Na^+Cl^-(s)

∴ +107 + 122 + 496 + (–349) + **E** = –411

Hence, the lattice energy of NaCl(s), **E** = **–787** kJ mol^{-1}

Definitions for enthalpy changes

The enthalpy changes involved in these two routes are shown below.

KEY POINT

The standard enthalpy change of formation ($\Delta H^\ominus_f$) is the enthalpy change that takes place when one mole of a compound in its standard state is formed from its constituent elements in their standard states under standard conditions.

Na(s) + ½Cl_2(g) ⟶ NaCl(s) $\Delta H^\ominus_f$ = –411 kJ mol^{-1}

Be careful when using $\Delta H^\ominus_{at}$ involving diatomic elements.

$\Delta H^\ominus_{at}$ relates to the formation of 1 mole of atoms. For chlorine, this involves ½Cl_2 only.

For gaseous molecules $\Delta H^\ominus_{at}$ can be determined from the **bond enthalpy** (B.E.).

Cl–Cl(g) ⟶ 2Cl(g)

$\Delta H^\ominus_{B.E.}$ = +244 kJ mol^{-1}

½Cl–Cl(g) ⟶ Cl(g)

$\Delta H^\ominus_{at}$ = +122 kJ mol^{-1}

KEY POINT

The standard enthalpy change of atomisation ($\Delta H^\ominus_{at}$) of an element is the enthalpy change that accompanies the formation of 1 mole of gaseous atoms from the element in its standard state.

Na(s) ⟶ Na(g) $\Delta H^\ominus_{at}$ = +107 kJ mol^{-1}

½Cl_2(g) ⟶ Cl(g) $\Delta H^\ominus_{at}$ = +122 kJ mol^{-1}

KEY POINT

The first ionisation energy ($\Delta H^\ominus_{I.E.}$) of an element is the enthalpy change that accompanies the removal of 1 electron from each atom in 1 mole of gaseous atoms to form 1 mole of gaseous 1+ ions.

Na(g) ⟶ Na^+(g) + e^- $\Delta H^\ominus_{I.E}$ = +496 kJ mol^{-1}

Be careful when using electron affinities. $\Delta H^\ominus_{E.A.}$ relates to the formation of 1 mole of 1– ions. For chlorine, this involves Cl(g) only.

KEY POINT

The first electron affinity ($\Delta H^\ominus_{E.A.}$) of an element is the enthalpy change that accompanies the addition of 1 electron to each atom in 1 mole of gaseous atoms to form 1 mole of gaseous 1– ions.

Cl(g) + e^- ⟶ Cl^-(g) $\Delta H^\ominus_{E.A.}$ = –349 kJ mol^{-1}

Factors affecting the size of lattice enthalpies

The strength of an ionic lattice and the value of its lattice enthalpy depend upon:

- ionic size
- ionic charge.

Effect of ionic size

The effect of increasing ionic size can be seen by comparing the lattice enthalpies of sodium halides as shown below.

Lattice energy has a negative value. You should use the term '*becomes less/more negative*' instead of '*becomes bigger/smaller*' to describe any trend in lattice energy.

compound	*lattice enthalpy / kJ mol^{-1}*	*ions*	*effect of size of halide ion*
NaCl	−787	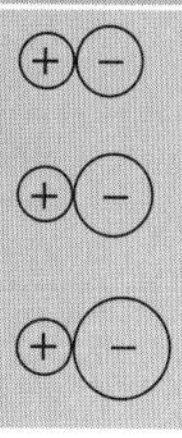	ionic size increases: • charge density decreases • attraction between ions decreases • lattice energy becomes less negative.
NaBr	−751		
NaI	−705		

Effect of ionic charge

The strongest ionic lattices with the most negative lattice enthalpies contain **small, highly charged ions**.

The diagram below compares the change in ionic size and ionic charge across Period 3 in the Periodic Table.

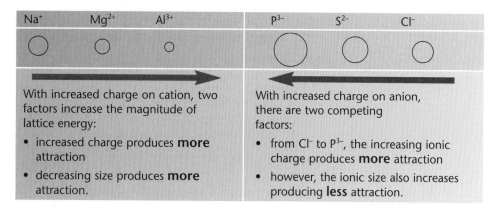

Na^+	Mg^{2+}	Al^{3+}	P^{3-}	S^{2-}	Cl^-

With increased charge on cation, two factors increase the magnitude of lattice energy: • increased charge produces **more** attraction • decreasing size produces **more** attraction.	With increased charge on anion, there are two competing factors: • from Cl^- to P^{3-}, the increasing ionic charge produces **more** attraction • however, the ionic size also increases producing **less** attraction.

Key points from AS

- **The thermal stability of s-block compounds**
  Revise AS page 67

Limitations of lattice enthalpies

Theoretical lattice enthalpies can be calculated by considering each ion as a perfect sphere. Any difference between this theoretical lattice enthalpy and lattice enthalpy calculated using a Born–Haber cycle indicates a degree of covalent bonding caused by polarisation. Polarisation is greatest between:

- a small densely charged cation and
- a large anion.

From the ions shown in the diagram above, we would expect the largest degree of covalency between Al^{3+} and P^{3-}.

Enthalpy change of solution

OCR M5
SALTERS M5

Key points from AS

- **Indirect determination of enthalpy changes**
  Revise AS pages 78–79

An ionic lattice dissolves in **polar** solvents (e.g. water). In this process, the giant ionic lattice is broken up by polar water molecules which surround each ion in solution.

When sodium chloride is dissolved in water, the enthalpy change of this process can be measured directly as the enthalpy change of solution.

The energy cycle

The complete energy cycle for dissolving sodium chloride in water is shown below. Definitions of these enthalpy changes are shown below.

Lattice enthalpy **forms** the ionic lattice. Notice that the energy change to break 1 mole of the ionic lattice = – (lattice enthalpy).

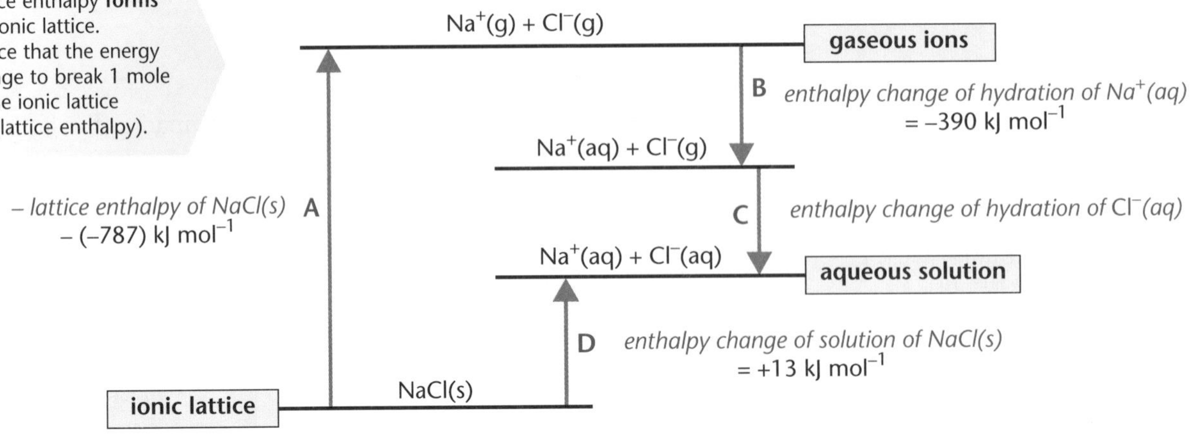

Route 1: A + B + C

Route 2: D

The production of hydrated ions comprises two changes:

- hydration of $Na^+(g)$
- hydration of $Cl^-(g)$.

Using Hess' Law: A + B + C = D

To determine the enthalpy change of hydration of $Cl^-(aq)$, **C**,

$$-\Delta H^{\ominus}_{L.E.}NaCl(s) + \Delta H^{\ominus}_{hyd}\ Na^+(g) + \mathbf{C} = \Delta H^{\ominus}_{solution}\ NaCl(s)$$

$$\therefore -(-787) + (-390) + \mathbf{C} = +13$$

Hence, the enthalpy change of hydration of $Cl^-(aq)$, **C = –384 kJ mol^{-1}**

Definitions for enthalpy changes

The enthalpy changes involved in these two routes are shown below.

KEY POINT

The **standard enthalpy change of solution** ($\Delta H^{\ominus}_{solution}$) is the enthalpy change that accompanies the dissolving of 1 mole of a solute in a solvent to form an infinitely dilute solution under standard conditions.

$NaCl(s) + aq \longrightarrow Na^+(aq) + Cl^-(aq)$ $\quad \Delta H^{\ominus}_{solution} = +13\ kJmol^{-1}$

KEY POINT

The **standard enthalpy change of hydration** ($\Delta H^{\ominus}_{hyd}$) of an ion is the enthalpy change that accompanies the hydration of 1 mole of gaseous ions to form 1 mole of hydrated ions in an infinitely dilute solution under standard conditions.

$Na^+(g) + aq \longrightarrow Na^+(aq)$ $\quad \Delta H^{\ominus}_{hyd} = -390\ kJmol^{-1}$

$Cl^-(g) + aq \longrightarrow Cl^-(aq)$ $\quad \Delta H^{\ominus}_{hyd} = -384\ kJmol^{-1}$

Factors affecting the magnitude of hydration enthalpy

The values obtained for each ion depend upon the size of the charge and the size of the ion.

ion	Na^+	Mg^{2+}	Al^{3+}	Cl^-	Br^-	I^-
$\Delta H^{\ominus}_{hyd}$/kJ mol^{-1}	–390	–1891	–4613	–384	–351	–307
effect of charge	→ • increasing ionic charge • greater attraction for water					
ionic radius/nm	0.102	0.072	0.053	0.180	0.195	0.215
effect of size	→ • decreasing ionic size • greater attraction for water			→ • increasing ionic size • less attraction for water		

Progress check

OCR

1 Using the information below:

(a) name each enthalpy change **A** to **E**

(b) construct a Born–Haber cycle for sodium bromide and calculate the lattice enthalpy (enthalpy change of lattice formation) of sodium bromide.

enthalpy change	*equation*	$\Delta H^{\ominus}$*/kJ mol*$^{-1}$
A	$Na(s) + \frac{1}{2}Br_2(l) \longrightarrow NaBr(s)$	–361
B	$Br(g) + e^- \longrightarrow Br^-(g)$	–325
C	$Na(s) \longrightarrow Na(g)$	+107
D	$Na(g) \longrightarrow Na^+(g) + e^-$	+496
E	$\frac{1}{2}Br_2(l) \longrightarrow Br(g)$	+112

OCR

SALTERS

2 You are provided with the following enthalpy changes.

$Na^+(g) + F^-(g) \longrightarrow NaF(s)$ $\quad \Delta H = -918$ kJ mol^{-1}

$Na^+(g) + aq \longrightarrow Na^+(aq)$ $\quad \Delta H = -390$ kJ mol^{-1}

$F^-(g) + aq \longrightarrow F^-(aq)$ $\quad \Delta H = -457$ kJ mol^{-1}

(a) Name each of these three enthalpy changes.

(b) Write an equation, including state symbols, for which the enthalpy change is the enthalpy change of solution of NaF.

(c) Calculate the enthalpy change of solution of NaF.

1 (a) A enthalpy change of formation of sodium bromide

B 1st electron affinity of bromine

C enthalpy change of atomisation of sodium

D 1st ionisation energy of sodium

E enthalpy change of atomisation of bromine

(b) –751 kJ mol^{-1}

2 (a) lattice enthalpy of NaF

enthalpy change of hydration of Na^+

enthalpy change of hydration of F^-

(b) $NaF(s) + aq \longrightarrow Na^+(aq) + F^-(aq)$

(c) +71 kJ mol^{-1}

Special properties of water arising from hydrogen bonding

SALTERS MS

When water changes state, the covalent bonds between the H and O atoms in an H_2O molecule are strong and do not break.

It is the intermolecular forces that break.

Hydrogen bonding is strong enough to have significant effects on physical properties, resulting in some unusual properties for water.

Water has an unexpectedly high boiling point

The boiling points of the Group 6 hydrides are shown below.

hydride	H_2O	H_2S	H_2Se	H_2Te
boiling point/K	373	212	232	271

The boiling points increase from $H_2S \longrightarrow H_2Te$:

- the number of electrons in molecules increase
- instantaneous dipole-induced dipole attractions increase
- more energy is required to break the intermolecular bonds to vaporise the hydrides.

Without hydrogen bonding, H_2O would be a gas at room temperature and pressure.

Enthalpy change of vaporisation is a measure of the energy required to convert water from a liquid to a gas.

But the **boiling point** of **H_2O** is **higher** than expected. This provides evidence that there are some extra forces acting between the molecules that must be broken to boil each hydride. These extra forces are hydrogen bonds.

This difference is also evident in the unusually high **enthalpy change of vaporisation** of water.

This has consequences for the Earth's climate.

- Energy is **required** to **vaporise** water molecules from the oceans into water vapour but this energy is **released** when water molecules **condense** as rain.
- Heat is transferred from the oceans to the land, stabilising the Earth's temperature across different climate zones.

Water has a relatively large specific heat capacity

Specific heat capacity, *c*, is a measure of the energy required to heat a substance.

Compared with similar compounds, water has an unusually high value specific heat capacity. Hydrogen bonds have to be overcome before energy can be transferred to water molecules to increase the kinetic energy of the molecules.

This has consequences for the Earth's climate.

- Because extra energy is needed to change the temperature of water, the vast stores of water in the oceans help to stabilise the Earth's temperature especially via ocean currents from one region to another.

Water is more dense than ice

Solids are usually denser than liquids – but ice is less dense than water.

- Solids are *usually* denser than liquids – but ice is less dense than water.
- Particles in solids are usually packed closer together than in liquids.
- Hydrogen bonds hold water molecules apart in an open lattice structure.

$\therefore$ ice is less dense than water.

The diagram below shows how the open lattice of ice collapses on melting.

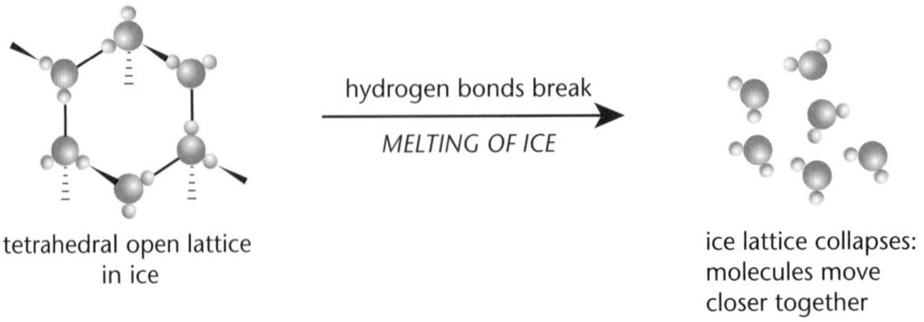

2.2 Entropy changes

After studying this section you should be able to:

- *explain that entropy is a measure of disorder*
- *calculate entropy changes from standard entropies*
- *understand that entropy, enthalpy and temperature determine the feasibility of a reaction*
- *apply rates and equilibrium to industrial applications*

LEARNING SUMMARY

Entropy

OCR MS
SALTERS MS

What is entropy?

Entropy, *S*, is a measure of disorder.

Entropy increases whenever particles become more disordered, e.g.

- when a gas spreads spontaneously through a room
- with increasing temperature
- during changes in state: solid ⟶ liquid ⟶ gas
- when a solid lattice dissolves
  $NaCl(aq) + aq \longrightarrow Na^+(aq) + Cl^-(aq)$
- in a reaction in which there is an increase in the number of gaseous molecules (i.e. when a gas is evolved)
  $NaHCO_3(s) + HCl(aq) \longrightarrow NaCl(aq) + H_2O(l) + CO_2(g)$
  $MgCO_3(s) \longrightarrow MgO(s) + CO_2(g)$

At 0 K, perfect crystals have zero entropy.

Calculating entropy changes

The standard entropy, $S^\ominus$, of a substance is the entropy content of one mole of a substance, under standard conditions.

The entropy change of a reaction can be calculated using standard entropies of the reactants and products.

$$\Delta S = \Sigma S^\ominus \text{ (products)} - \Sigma S^\ominus \text{ (reactants)}$$

KEY POINT

Example

The equation below is for the carbon reduction of chromium oxide, Cr_2O_3.

$$Cr_2O_3(s) + 3C(s) \longrightarrow 2Cr(s) + 3CO(g)$$

Notice that standard entropies have units of J K^{-1} mol^{-1}.

	$Cr_2O_3(s)$	C(s)	Cr(s)	CO(g)
$S^\ominus$/J K^{-1} mol^{-1}	+81	+6	+24	+198

$\Delta S = \Sigma S^\ominus$ (products) – $\Sigma S^\ominus$ (reactants)

= [(2 x 24) + (3 x 198)] – [(+81) + (3 x 6)] = 543 J K^{-1} mol^{-1}

Spontaneous processes

A **spontaneous** process is one that proceeds of its **own accord**.

Processes lead to **increased stability** if they lead to a **lower energy** state.

This is obviously the situation in an exothermic reaction in which the enthalpy (heat energy) decreases during a reaction. It therefore seems surprising that **some** reactions are **endothermic** and that these can **occur spontaneously** at room temperature.

Enthalpy changes alone cannot explain whether a reaction takes place.

For this, we must consider **both** the **ΔH and ΔS**.

Free energy

OCR MS

Whether a chemical or physical process takes place is dependent on

- the temperature, T,
- the entropy change in the system, ΔS,
- the enthalpy change, ΔH, with the surroundings.

The change in *Free Energy*, ΔG, automatically determines the balance between enthalpy change, entropy change and temperature:

$$\Delta G = \Delta H - T\Delta S$$

The balance between entropy and enthalpy is the feasibility of a reaction.

Feasibility of a reaction

Processes are spontaneous (feasible) when ΔG is negative. The sign of ΔG depends upon ΔH and ΔS, as shown below.

ΔH	ΔS	ΔG
–ve	+ve	always –ve: reaction feasible
+ve	–ve	always +ve: reaction never feasible
–ve	–ve	–ve at low temperatures
+ve	+ve	–ve at high temperatures

ΔG varies with temperature: when ΔH and ΔS have the same sign, the feasibility of the process depends on the temperature, T.

Be careful with units:

ΔH: kJ mol^{-1}

ΔS: J K^{-1} mol^{-1}

- $\Delta G = \Delta H - T\Delta S$
- The reaction is feasible when $\Delta G < 0$.

Example

The equation below is for the carbon reduction of chromium oxide, Cr_2O_3.

$Cr_2O_3(s) + 3C(s) \longrightarrow 2Cr(s) + 3CO(g)$

$\Delta H = +807$ kJ mol^{-1}; $\Delta S = 543$ J K^{-1} mol^{-1}

What is the minimum temperature that this reaction takes place spontaneously?

$\Delta S = 543$ J K^{-1} mol^{-1} = 0.543 kJ K^{-1} mol^{-1}

$\Delta G = \Delta H - T\Delta S$;

Minimum temperature when $\Delta G = 0$.

$\therefore\ 0 = \Delta H - T\Delta S;\quad T = \dfrac{\Delta H}{\Delta S} = \dfrac{807}{0.543} = 1486\text{ K}$

Total entropy

SALTERS MS

For a chemical process to take place, there are two types of entropy (disorder) that need to be considered:

- the entropy change of the chemical system, ΔS_{system}
- the entropy change of the surroundings, $\Delta S_{surroundings}$

ΔS_{system}

This is the disorder of the particles making up the chemicals – the chemical 'system'. ΔS_{system} measures the number of ways that the particles can be arranged. It also measures the number of ways that energy can be distributed between the particles.

Entropy is a measure of the number of ways that molecules and their associated energy quanta can be arranged.

ΔS_{system} can be calculated using:

$$\Delta S_{system} = \Sigma S^{\ominus}\text{ (products)} - \Sigma S^{\ominus}\text{ (reactants)}$$

A system becomes more stable when ΔS_{system} is positive.

$\Delta S_{surroundings}$

This is the disorder of energy in the surroundings.

In terms of energy, the surroundings become more stable when $\Delta S_{surroundings}$ increases and energy spreads from the chemicals into the surroundings in an exothermic process.

$\Delta S_{surroundings}$ is related to the enthalpy change and temperature, in K:

$$\Delta S_{surroundings} = -\frac{\Delta H}{T}$$

The surrounding system becomes more stable during an exothermic reaction when $\Delta S_{surroundings}$ is positive.

Feasibility of a reaction

ΔS_{total} only tells us the 'thermodynamic stability'. Even if a reaction is thermodynamically stable, it may not seem to take place if the activation energy is great and the rate of reaction is slow ('kinetic stability').

The total entropy, ΔS_{total}, is the overall balance of entropy from the system and from the surroundings:

$$\Delta S_{total} = \Delta S_{system} + \Delta S_{surroundings}$$

or

$$\Delta S_{total} = \Delta S_{system} - \frac{\Delta H}{T}$$

For a process to take place, the natural direction of change is towards increasing total entropy when ΔS_{total} is positive, i.e. $\Delta S_{total} > 0$.

The process is then feasible and takes place 'spontaneously'.

Example

The equation below is for the carbon reduction of chromium oxide, Cr_2O_3.

$Cr_2O_3(s) + 3C(s) \longrightarrow 2Cr(s) + 3CO(g)$

$\Delta H = +807\text{ kJ mol}^{-1}$; $\Delta S = 543\text{ J K}^{-1}\text{ mol}^{-1}$

What is the minimum temperature that this reaction takes place at spontaneously?

Be careful with units:
ΔH: kJ mol^{-1}
ΔS: J K^{-1} mol^{-1}

$\Delta S_{system} = 543\text{ J K}^{-1}\text{ mol}^{-1} = 0.543\text{ kJ K}^{-1}\text{ mol}^{-1}$

$$\Delta S_{total} = \Delta S_{system} - \frac{\Delta H}{T}$$

Minimum temperature when $\Delta S_{total} = 0$.

$$\therefore\ \Delta S_{system} - \frac{\Delta H}{T} = 0;\ \Delta S_{system} = \frac{\Delta H}{T};\ T = \frac{\Delta H}{\Delta S_{system}} = \frac{807}{0.543} = 1486\text{ K}$$

Progress check

1 State whether the equations below have a positive or negative entropy change.

(a) $H_2O(g) \longrightarrow H_2O(l)$

(b) $N_2(g) + 3H_2(g) \longrightarrow 2NH_3(g)$

(c) $CaCl_2(s) + aq \longrightarrow Ca^{2+}(aq) + 2Cl^-(aq)$

2 The equation below is for the thermal decomposition of calcium carbonate.

$CaCO_3(s) \longrightarrow CaO(s) + CO_2(g)$ $\Delta H = +178$ kJ mol^{-1}

	$CaCO_3(s)$	$CaO(s)$	$CO_2(g)$
$S^{\ominus}$/ J K^{-1} mol^{-1}	+93	+40	+214

(a) What is ΔS?

(b) What is the minimum temperature for decomposition to take place?

1 (a) –ve; (b) –ve; (c) +ve.
2 (a) 161 J K^{-1} mol^{-1}; (b) 1106 K

Applying rates and equilibrium to industrial processes

OCR M5
SALTERS M5

Industrial processes need to obtain economic yields of the product being manufactured. In the section above, we discussed the importance of the balance between **enthalpy, entropy** and **temperature**.

Equilibrium

You do not need to know the actual industrial conditions used in the Haber process.

We can arrive at these conclusions by using le Chatelier's principle from AS Chemistry.

In the Haber process for the production of ammonia

$N_2(g) + 3H_2(g) \rightleftharpoons 2NH_3(g)$ $\Delta H^{\ominus} = -92$ kJ mol^{-1}

- the right-hand side has **fewer moles of gas** than the left-hand side (reactants), favoured by a **high** pressure
- the forward reaction is **exothermic**, favoured by a **low** temperature.

The optimum equilibrium conditions for maximum equilibrium yield are:
high pressure and low temperature.

The need for compromise

The rate of a reaction increases with higher temperatures and pressures – but temperature has a much larger effect.

In reality it may not be feasible to use optimum equilibrium conditions.

- Compressing gases to **high pressures** has a high **energy cost**. There are also considerable **safety** implications of using very high pressures.
- At **low temperatures**, the rate of reaction is **slow** as few molecules possess the necessary activation energy of the reaction.

There needs to be a compromise.

Compromising
Optimum conditions, reaction rate, feasibility, reality, safety and economics must be considered.

- Use of a **high enough pressure** to **increase** the **equilibrium yield** without increasing **energy costs** in generating the pressure or compromising **safety**.
- Increase the temperature just enough to increase the rate of reaction so that the reaction takes place in a realistic time frame, whilst still producing an **adequate equilibrium yield**.

Equilibrium is never reached

Another factor is that industrial processes do not operate at equilibrium:

- products are removed – we never have a closed system
- there is insufficient time for equilibrium to be reached.

But it is still important to consider the conditions needed to secure a high equilibrium yield.

The search for better processes

Hazards can be reduced by developing processes that reduce risks from toxic reactants, explosions, acidic or flammable gases, toxic emissions, etc.

The chemical industry takes steps to maximise the **atom economy** of the process. Many factors need to be considered, including:

- **recycling** unreacted reagents
- developing **alternative reactions** and processes
- using **more available** raw materials, preferably **renewable**
- using **catalysts** to allow processes to occur at **lower temperatures**, cutting energy needs
- developing **simpler processes** requiring less costly chemical plants
- using **co-products** and **by-products**, rather than disposing of them.

Global warming from energy production

SALTERS MS

Key points from AS

- **Global warming**
  Revise AS pages 123–124

Much of the CO_2 released into the atmosphere dissolves in the oceans.

The concentration of carbon dioxide in the atmosphere depends on natural processes including:

- photosynthesis
- respiration
- the solubility of the gas in surface waters.

Combustion of non-renewable hydrocarbon fuels for energy is causing an increase in the atmospheric concentration of carbon dioxide. Carbon dioxide is a greenhouse gas, implicated as a likely cause for global warming.

The **greenhouse effect** of a gas (e.g. CO_2) depends on:

- its atmospheric concentration
- its ability to absorb infra-red radiation.

Reducing atmospheric CO_2 levels

Various approaches are being considered to reduce atmospheric CO_2 levels. These include:

- more economical use of fuels
- the use of alternative fuels (including hydrogen, see pages 57–59)
- capture and storage of CO_2
- increased photosynthesis.

2.3 Electrochemical cells

After studying this section you should be able to:

- *understand the principles of an electrochemical cell*
- *define the term 'standard electrode (redox) potential', $E^{\ominus}$*
- *describe how to measure standard electrode potentials*
- *calculate a standard cell potential from standard electrode potentials*
- *describe applications of electrochemical cells, including fuel cells*
- *understand that rusting is an electrochemical process*

LEARNING SUMMARY

Key points from AS

- **Redox reactions**
  Revise AS pages 35–37

During the study of redox in AS Chemistry you learnt:

- that oxidation and reduction involve transfer of electrons in a redox reaction
- the rules for assigning oxidation states
- how to construct an overall redox equation from two half-equations.

These key principles are built upon by considering electrochemical cells in which redox reactions provide electrical energy.

Electricity from chemical reactions

OCR M5
SALTERS M4

In a redox reaction, electrons are transferred between the reacting chemicals.

For example, zinc reacts with aqueous copper(II) ions in a redox reaction.

$$Zn(s) + Cu^{2+}(aq) \longrightarrow Zn^{2+}(aq) + Cu(s)$$

oxidation: $Zn(s) \longrightarrow Zn^{2+}(aq) + 2e^-$

reduction: $Cu^{2+}(aq) + 2e^- \longrightarrow Cu(s)$

Electrochemical cells are the basis of all batteries. They are used as a portable form of electricity.

If this reaction is carried out simply by mixing zinc with aqueous copper(II) ions:

- heat energy is produced and, unless used as useful heat, is often wasted.

An electrochemical cell controls the transfer of electrons:

- electrical energy is produced and can be used for electricity.

The zinc-copper electrochemical cell

An electrochemical cell comprises of two **half-cells**. In each half-cell, the **oxidation** and **reduction** processes take place **separately**.

A half-cell contains the oxidised and reduced species from a half-equation.

The species can be converted into one another by addition or removal of electrons.

A zinc-copper cell has two half-cells:

- an oxidation half-cell containing reactants and products from the oxidation half-reaction: $Zn^{2+}(aq)$ and Zn(s)
- a reduction half-cell containing reactants and products from the reduction half-reaction: $Cu^{2+}(aq)$ and Cu(s).

The circuit is completed using:

- a connecting **wire** which transfers **electrons** between the half-cells
- a **salt bridge** which transfers **ions** between the half-cells.

The salt bridge consists of filter paper soaked in aqueous KNO_3 or NH_4NO_3.

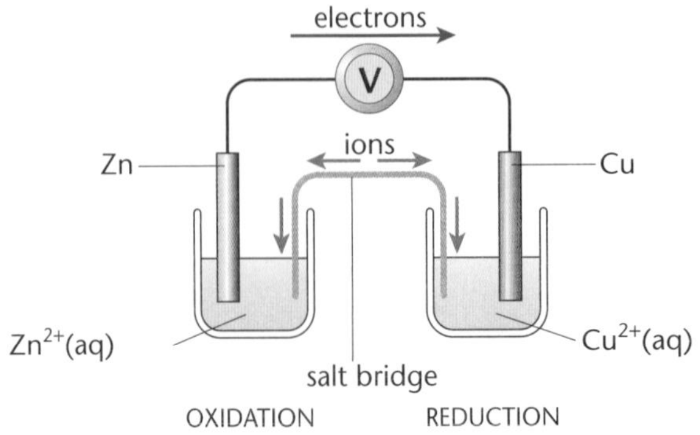

Remember these charge carriers:

ELECTRONS through the wire;

IONS through the salt bridge.

oxidation half-cell: Zn^{2+}(aq) and Zn(s)
- **Zn is oxidised** to Zn^{2+}: Zn(s) ⟶ Zn^{2+}(aq) + $2e^-$
- electrons are **supplied** to the external circuit
- the polarity is **negative**.

reduction half-cell: Cu^{2+}(aq) and Cu(s)
- **Cu^{2+} is reduced** to Cu: Cu^{2+}(aq) + $2e^-$ ⟶ Cu(s)
- electrons are **taken** from the external circuit
- the polarity is **positive**.

The measured cell e.m.f. of 1.10 V indicates that there is a potential difference of 1.10 V between the copper and zinc half-cells.

Standard electrode potentials

OCR M5
SALTERS M4

A **hydrogen half-cell** is used as the standard for the measurement of standard electrode potentials.

The standard hydrogen half-cell is based upon the half-reaction below.

$$H^+(aq) + e^- \rightleftharpoons \tfrac{1}{2}H_2(g)$$

A hydrogen half-cell typically comprises
- 1 mol dm^{-3} hydrochloric acid as the source of H^+(aq)
- a supply of hydrogen gas, H_2(g) at 100 kPa
- an inert platinum electrode on which the half-reaction above takes place.

KEY POINT

The **standard electrode potential** of a half-cell, $E^{\ominus}$, is the e.m.f. of a half-cell compared with a standard hydrogen half-cell.
All measurements are at **298 K** with solution **concentrations of 1 mol dm^{-3}** and gas **pressures of 100 kPa**.

Measuring standard electrode potentials

Half-cells comprising a metal and a metal ion

The diagram below shows how a hydrogen half-cell can be used to measure the standard electrode potential of a copper half-cell.

A high-resistance voltmeter is used to minimise the current that flows.

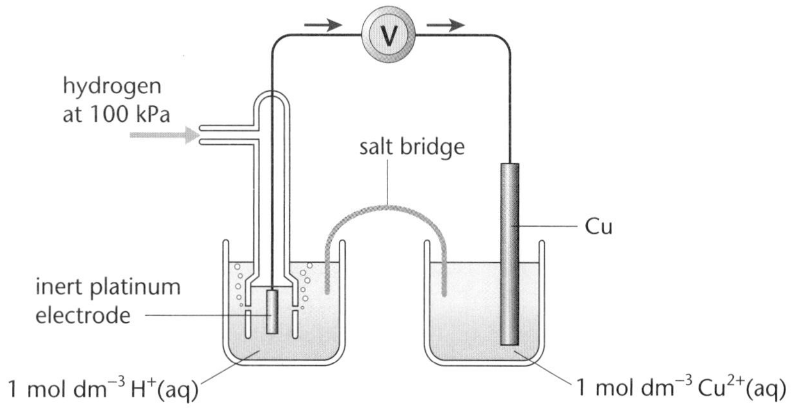

Hydrogen half-cell

KEY POINT

- The standard electrode potential of a half-cell represents its contribution to the cell e.m.f.
- The contribution made by the hydrogen half-cell to the cell e.m.f. is defined as 0 V.
- The **sign** of the standard electrode potential of a half-cell indicates its **polarity** compared with the hydrogen half-cell.

As with the hydrogen half-cell, the platinum electrode provides a surface on which the half-reaction can take place.

Half-cells comprising ions of different oxidation states

The half-reaction in a half-cell can be between aqueous ions of the same element with different oxidation states. To allow electrons to pass into the half-cell, an inert electrode of platinum is used.

A half-cell can contain $Fe^{3+}(aq)$ and $Fe^{2+}(aq)$ ions:

$$Fe^{3+}(aq) \rightleftharpoons Fe^{2+}(aq) + e^-$$

A **standard** $Fe^{3+}(aq)/Fe^{2+}(aq)$ half-cell requires:

- a solution containing **both**
  - 1 mol dm^{-3} $Fe^{3+}(aq)$ **and**
  - 1 mol dm^{-3} $Fe^{2+}(aq)$
- an inert platinum electrode with a connecting wire.

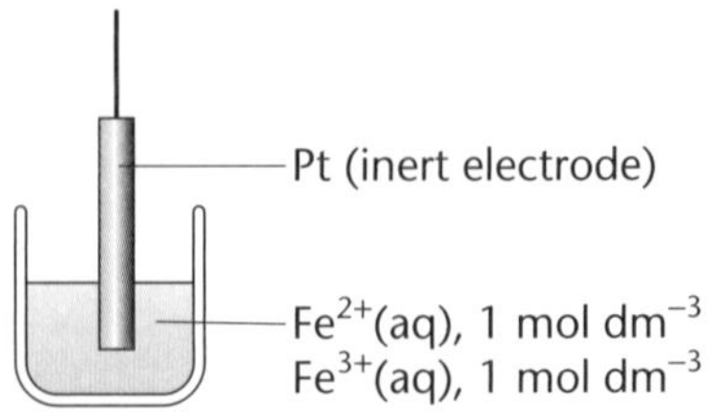

Alternatively the solution can contain **equal** concentrations of Fe^{2+} and Fe^{3+}.

The electrochemical series

OCR M5
SALTERS M4

A standard electrode potential indicates the availability of electrons from a half-reaction.

Metals are reducing agents.

- Metals typically react by donating electrons in a redox reaction.
- Reactive metals have the greatest tendency to donate electrons and their half-cells have the most negative electrode potentials.

Non-metals are oxidising agents.

- Non-metals typically react by accepting electrons in a redox reaction.
- Reactive non-metals have the greatest tendency to accept electrons and their half-cells have the most positive electrode potentials.

Key points from AS

- **Reactivity of s-block elements**
  Revise AS page 66
- **The relative reactivity of the halogens as oxidising agents**
  Revise AS pages 68–69

The **electrochemical series** shows electrode reactions listed in order of their $E^{\ominus}$ values. The electrochemical series below lists with the **most positive** $E^{\ominus}$ value at the top. This enables oxidising and reducing agents to be easily compared. The reduced form is shown on the right-hand side of the electrode equilibrium.

strongest oxidising agent ↑

electrode reaction					$E^{\ominus}$ / V
$F_2(g)$	+	$2e^-$	⇌	$2F^-(aq)$	+2.87
$Cl_2(g)$	+	$2e^-$	⇌	$2Cl^-(aq)$	+1.36
$Br_2(l)$	+	$2e^-$	⇌	$2Br^-(aq)$	+1.09
$Ag^+(aq)$	+	e^-	⇌	$Ag(s)$	+0.80
$Cu^{2+}(aq)$	+	$2e^-$	⇌	$Cu(s)$	+0.34
$H^+(aq)$	+	e^-	⇌	$\frac{1}{2}H_2(g)$	0
$Fe^{2+}(aq)$	+	$2e^-$	⇌	$Fe(s)$	−0.44
$Zn^{2+}(aq)$	+	$2e^-$	⇌	$Zn(s)$	−0.76
$Cr^{3+}(aq)$	+	$3e^-$	⇌	$Cr(s)$	−0.77
$K^+(aq)$	+	e^-	⇌	$K(s)$	−2.92

↓ *strongest reducing agent*

This version of the electrochemical series shows standard electrode potentials listed with the **most positive** $E^{\ominus}$ value at the top.

You will find versions of the electrochemical series listed with the **most negative** $E^{\ominus}$ value at the top.

This does not matter. The important point is that the potentials are listed **in order** and **can be compared**.

KEY POINT

In the electrochemical series:
- the **oxidised** form is on the **left-hand side** of the half-equation
- the **reduced** form is on the **right-hand side** of the half-equation.

Standard cell potentials and cell reactions

OCR M5
SALTERS M4

The **standard cell potential** of a cell is the e.m.f. acting between the two half-cells making up the cell under standard conditions.

The **cell reaction** is:

- the overall process taking place in the cell
- the sum of the reduction and oxidation half-reactions taking place in each half-cell.

Key points from AS

- **Combining half-equations**
  Revise AS page 37

Calculating the standard cell potential of a silver-iron(II) cell

Identify the two relevant half-reactions and the polarity of each electrode.

The more positive of the two systems is the positive terminal of the cell.

$Ag^{+}(aq) + e^{-} \rightleftharpoons Ag(s)$ $E^{\ominus} = +0.80$ V *positive terminal*

$Fe^{2+}(aq) + 2e^{-} \rightleftharpoons Fe(s)$ $E^{\ominus} = -0.44$ V *negative terminal*

$E^{\ominus}_{cell}$ is the difference between the standard electrode potentials.

The standard electrode potential is the difference between the $E^{\ominus}$ values:

Simply subtract the $E^{\ominus}$ of the negative terminal from the $E^{\ominus}$ of the positive terminal:

$$E^{\ominus}_{cell} = E^{\ominus}\ (\textit{positive terminal}) - E^{\ominus}\ (\textit{negative terminal})$$

$$\therefore\ E^{\ominus}_{cell} = 0.080 - (-0.44) = \mathbf{1.24\ V}$$

Determination of the cell reaction in a silver-iron(II) cell

The more negative half-equation is reversed. The negative terminal is then on the left-hand side.

Work out the actual direction of the half-equations.

The **half-equation** with the **more negative $E^{\ominus}$** value provides the electrons and proceeds to the left. This half-equation is reversed so that the two half-reactions taking place can be clearly seen and compared.

$Ag^{+}(aq) + e^{-} \longrightarrow Ag(s)$ *reaction at positive terminal*

$Fe(s) \longrightarrow Fe(aq) + 2e^{-}$ *reaction at negative terminal*

This is discussed in detail in the AS Chemistry Study Guide, page 37.

The overall cell reaction can be found by adding the half-equations.

The silver half-equation must first be multiplied by '2' to balance the electrons.

The two half-equations are then added.

$$2Ag^{+}(aq) + Fe(s) \longrightarrow 2Ag(s) + Fe^{2+}(aq)$$

Electrochemical cells

OCR M5
SALTERS M5

Electrochemical cells can be used as a commercial source of electrical energy.

Cells can be divided into three main types:

- non-rechargeable cells – the cell is used until the chemicals are used up to such a point at which the voltage falls
- rechargeable cells – the cell reaction can readily be reversed when recharging, allowing the cell to be used again
- fuel cells – the cell reaction requires external supplies of a fuel and an oxidant, which are consumed and need to be replenished.

Other common examples include:

- nickel and cadmium (NiCad) batteries used in rechargeable batteries
- lithium 'button' cells used in watches
- lithium-ion batteries used in laptops.

All these cells work on the same principle, using two redox systems.

An electrochemical cell comprises two half-cells with different electrode potentials. For example, a simple cell can be set up based on zinc and copper.

The redox systems are shown below.

$Zn^{2+}(aq) + 2e^{-} \rightleftharpoons Zn(s)$ $E^{\ominus} = -0.74$ V

$Cu^{2+}(aq) + 2e^{-} \rightleftharpoons Cu(s)$ $E^{\ominus} = +0.34$ V

The more negative zinc system provides the electrons:

Overall: $Zn(s) + Cu^{2+}(aq) \longrightarrow Cu(s) + Zn^{2+}(aq)$

Chargeable and non-rechargeable cells are used as our modern-day cells and batteries.

Fuel cells for energy

OCR M5

In a fuel cell, energy from the reaction of a fuel with oxygen is used to create a voltage.

- The reactants flow in and products flow out, while the electrolyte remains in the cell.
- Fuel cells can operate virtually continuously as long as the fuel and oxygen continue to flow into the cell. Fuel cells do not have to be recharged.

In the future, a 'hydrogen economy' based on hydrogen fuel cells may well contribute greatly to energy needs.

The diagram below shows a simple hydrogen–oxygen fuel cell.

- At the negative electrode (cathode),

  $H_2(g) \longrightarrow 2H^+(aq) + 2e^-$

- At the positive electrode (anode),

  $\frac{1}{2}O_2(g) + 2H^+(aq) + 2e^- \longrightarrow H_2O(l)$

- Overall: $H_2(g) + \frac{1}{2}O_2(g) \longrightarrow H_2O(l)$

In exams, you are unlikely to have to recall a specific storage cell.

But do take care! Your specification **may** expect you to know the hydrogen fuel cell.

You might also be expected to make predictions on a supplied electrochemical cell. All relevant electrode potentials and other data would be supplied though.

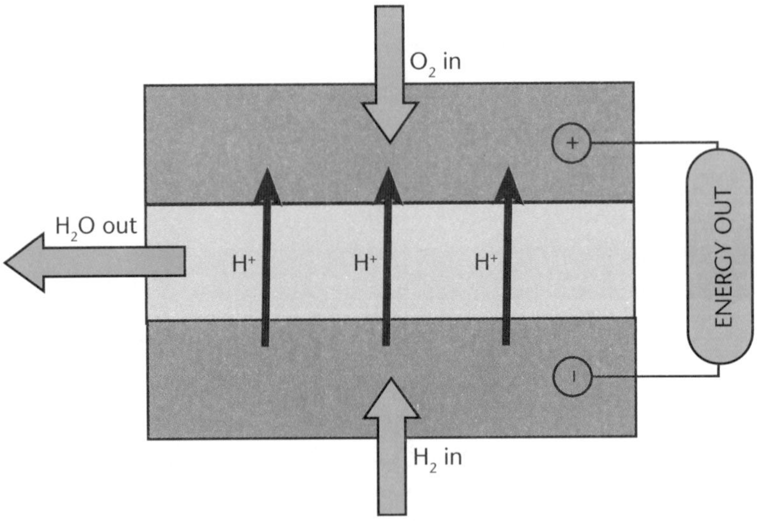

Other hydrogen fuel cells operate with an alkaline electrolyte.

Development of fuel cell vehicles (FCVs)

Fuel cells are being developed as an alternative to finite oil-based fuels in cars.

Scientists in the car industry are developing fuel cell vehicles (FCVs), fuelled by:

- hydrogen gas
- hydrogen-rich fuels.

Hydrogen-rich fuels include methanol, natural gas, or petrol. These are converted into hydrogen gas by an onboard 'reformer'. These fuels are easier to transport and deliver into an FCV than hydrogen gas.

FCVs have advantages over conventional petrol or diesel-powered vehicles.

- Pure hydrogen emits only water.
- Hydrogen-rich fuels produce only small amounts of air pollutants and CO_2.
- Efficiency can be more than twice that of similarly-sized conventional vehicles. Other advanced technologies are being investigated to further increase efficiency.

Advantages of hydrogen fuel cells:

less pollution and less CO_2,

greater efficiency.

Hydrogen might be stored in FCVs:

- as a liquid under pressure
- adsorbed on the surface of a solid material
- absorbed within a solid material.

Liquefying hydrogen is costly in terms of energy use.

Safety is a major concern: Hydrogen is very flammable.

Hydrogen is an 'energy carrier' and not an 'energy source'. Hydrogen is first made either by electrolysis of water or by reacting methanol (a finite fuel) with steam.

Limitations of hydrogen fuel cells

There are logistical problems in the development of hydrogen fuel cells:

- storing and transporting hydrogen, in terms of safety, feasibility of a pressurised liquid and a limited life cycle of a solid 'adsorber' or 'absorber'
- limited lifetime (requiring regular replacement and disposal) and high production costs
- use of toxic chemicals in their production
- initial manufacture of hydrogen requires energy
- public and political acceptance of hydrogen as a fuel.

Future energy needs

Political and social desire to move to a hydrogen economy has many obstacles. Many people do not realise that energy is needed to produce hydrogen and that fuel cells have a finite life.

Progress check

1 Use the standard electrode potentials on page 56, to answer the questions that follow.

(a) Calculate the standard cell potential for the following cells.

(i) F_2/F^- and Ag^+/Ag

(ii) Ag^+/Ag and Cr^{3+}/Cr

(iii) Cu^{2+}/Cu and Cr^{3+}/Cr.

(b) For each cell in (a) determine which half-reactions proceed, and hence construct, the cell reaction.

1 (a) (i) 2.07 V (ii) 1.57 V (iii) 1.11 V.

(b) $F_2 + 2Ag \longrightarrow 2F^- + 2Ag^+$

$3Ag^+ + Cr \longrightarrow 3Ag + Cr^{3+}$

$3Cu^{2+} + 2Cr \longrightarrow 3Cu + 2Cr^{3+}$

Corrosion as an electrochemical process

SALTERS M5

Rusting

Rusting is an electrochemical process involving **iron**, **oxygen** and **water**.

The key redox systems involved are shown below:

$$Fe^{2+} + 2e^- \rightleftharpoons Fe \qquad E^\ominus = -0.44\ V$$

$$\tfrac{1}{2}O_2 + H_2O + 2e^- \rightleftharpoons 2OH^- \qquad E^\ominus = +0.44\ V$$

- The **Fe^{2+}/Fe** system has the **more negative** $E^\ominus$ value and its half-reaction will **supply electrons**.
- This equation is therefore reversed.

oxidation:	Fe	$\longrightarrow$	$Fe^{2+} + 2e^-$
reduction:	$\tfrac{1}{2}O_2 + H_2O + 2e^-$	$\longrightarrow$	$2OH^-$
overall:	$Fe + \tfrac{1}{2}O_2 + H_2O$	$\longrightarrow$	$Fe^{2+} + 2OH^-$

- The Fe^{2+} and OH^- ions now combine to form iron(II) hydroxide.
- Finally $Fe(OH)_2$ is **oxidised** by **oxygen** in the presence of **water**, forming **rust**:

$$Fe(OH)_2 \xrightarrow{O_2/H_2O} \underset{\text{RUST}}{Fe_2O_3{\bullet}xH_2O}$$

Corrosion prevention

The main methods for protecting iron and steel from corrosion are:

- sacrificial protection by galvanising and use of zinc blocks
- barrier protection using oil, grease, paint or a polymer coating.

In **sacrificial protection**,

- the iron is in contact with a metal with a more negative $E^\ominus$ value
- Mg or Zn are usually used.

Sacrificial protection is used to protect cars (a zinc coating) and North Sea oil platforms (zinc blocks).

$$Mg^{2+} + 2e^- \rightleftharpoons Mg \qquad E^\ominus = -2.36\ V$$

$$Zn^{2+} + 2e^- \rightleftharpoons Zn \qquad E^\ominus = -0.76\ V$$

$$Fe^{2+} + 2e^- \rightleftharpoons Fe \qquad E^\ominus = -0.44\ V$$

more negative $E^\ominus$

When an electrochemical cell is set up the iron is protected:

- the **metal** with a **more negative** $E^\ominus$ value **supplies the electrons** and is sacrificed.

Once reacted, the Zn or Mg can be replaced but the iron is still there.

Recycling of scrap iron and steel

All steel packaging except aerosols can be recycled.

- Incineration is used to clean the metal.
- Iron and steel are separated using magnets.
- The composition of new steel is easily adjusted.
- Scrap is used to adjust the temperature of the furnace.

2.4 Predicting redox reactions

After studying this section you should be able to:

LEARNING SUMMARY

- *use standard cell potentials to predict the feasibility of a reaction*
- *recognise the limitations of such predictions*
- *understand redox processes in the nitrogen cycle*

Key points from AS

- **Redox reactions**
  Revise AS pages 35–37

During AS Chemistry, you learnt how to combine two half-equations to construct a full redox equation.

Standard electrode potentials can be used to predict the direction of any aqueous redox reaction from the relevant half-reactions.

Using standard electrode potentials to make predictions

OCR M5
SALTERS M4

By studying standard electrode potentials, predictions can be made about the likely feasibility of a reaction.

The table below shows three redox systems.

	redox system	$E^{\ominus}$ /V
A	$H_2O_2(aq) + 2H^+ + 2e^- \rightleftharpoons 2H_2O(l)$	+1.77
B	$I_2(aq) + 2e^- \rightleftharpoons 2I^-(aq)$	+0.54
C	$Fe^{3+}(aq) + 3e^- \rightleftharpoons Fe(s)$	–0.04

We can predict the redox reactions that may take place between these species by treating the redox systems as if they are half-cells.

Notice that a reaction takes place between reactants from different sides of each half equation.

Comparing the redox systems A and B

- The I_2/I^- system has the less positive $E^{\ominus}$ value and its half-reaction will proceed to the left, donating electrons:

$$\mathbf{H_2O_2}(aq) + \mathbf{2H^+}(aq) + 2e^- \longrightarrow 2H_2O(l) \qquad E^{\ominus} = +1.77\text{ V}$$
$$I_2(aq) + 2e^- \longleftarrow \mathbf{2I^-}(aq) \qquad E^{\ominus} = +0.54\text{ V}$$

We would therefore predict that $\mathbf{H_2O_2}$ **and** $\mathbf{H^+}$ would react with $\mathbf{I^-}$.

Comparing the redox systems B and C

- The Fe^{3+}/Fe system has the more negative $E^{\ominus}$ value and its half-reaction will proceed to the left, donating electrons:

$$\mathbf{I_2}(aq) + 2e^- \longrightarrow 2I^-(aq) \qquad E^{\ominus} = +0.54\text{ V}$$
$$Fe^{3+}(aq) + 3e^- \longleftarrow \mathbf{Fe}(s) \qquad E^{\ominus} = -0.04\text{ V}$$

We would therefore predict that $\mathbf{I_2}$ would react with **Fe**.

Comparing the redox systems A and C

- The Fe^{3+}/Fe system has the more negative $E^{\ominus}$ value and its half-reaction will proceed to the left, donating electrons:

$$\mathbf{H_2O_2}(aq) + \mathbf{2H^+}(aq) + 2e^- \longrightarrow 2H_2O(l) \qquad E^{\ominus} = +1.77\text{ V}$$
$$Fe^{3+}(aq) + 3e^- \longleftarrow \mathbf{Fe}(s) \qquad E^{\ominus} = -0.04\text{ V}$$

We would therefore predict that $\mathbf{H_2O_2}$ **and** $\mathbf{H^+}$ would react with **Fe**.

KEY POINT

When comparing two redox systems, we can predict that a reaction may take place between:

- the stronger reducing agent with the more negative $E^{\ominus}$, on the right-hand side of the redox system
- the stronger oxidising agent with the more positive $E^{\ominus}$, on the left-hand side of the redox system.

Limitations of standard electrode potentials

OCR M5
SALTERS M4

A change in electrode potential resulting from concentration changes means that predictions made on the basis of the **standard** value may not be valid.

Electrode potentials and ionic concentration

Non-standard conditions alter the value of an electrode potential.

The half-equation for the copper half-cell is shown below.

$$Cu^{2+}(aq) + 2e^- \rightleftharpoons Cu(s)$$

Using le Chatelier's principle, by increasing the concentration of $Cu^{2+}(aq)$:

- the equilibrium opposes the change by moving to the right
- electrons are removed from the equilibrium system
- the electrode potential becomes more positive.

Will a reaction actually take place?

Remember that these are equilibrium processes.

- Predictions can be made but these give no indication of the reaction rate which may be extremely slow, caused by a large activation energy.
- The actual conditions used may be different from the standard conditions used to record $E^{\ominus}$ values. This will affect the value of the electrode potential (see above).
- Standard electrode potentials apply to aqueous equilibria. Many reactions take place that are not aqueous.

As a general working rule:

- the larger the difference between $E^{\ominus}$ values, the more likely that a reaction will take place
- if the difference between $E^{\ominus}$ values is less than 0.4 V, then a reaction is unlikely to go to completion.

Progress check

1 For each of the three predicted reactions on page 61, construct full redox equations.

2 Use the standard electrode potentials below to answer the questions that follow.

$Fe^{3+}(aq) + e^-$	$\rightleftharpoons$	$Fe^{2+}(aq)$	$E^{\ominus} = +0.77$ V
$Br_2(l) + 2e^-$	$\rightleftharpoons$	$2Br^-(aq)$	$E^{\ominus} = +1.07$ V
$Ni^{2+}(aq) + 2e^-$	$\rightleftharpoons$	$Ni(s)$	$E^{\ominus} = -0.25$ V
$O_2(g) + 4H^+(aq) + 4e^-$	$\rightleftharpoons$	$2H_2O(l)$	$E^{\ominus} = +1.23$ V

(a) Arrange the reducing agents in order with the most powerful first.

(b) Predict the reactions that could take place.

1 **A** and **B**: $H_2O_2(aq) + 2H^+(aq) + 2I^-(aq) \longrightarrow I_2(aq) + 2H_2O(l)$; **B** and **C**: $2Fe(s) + 3I_2(aq) \longrightarrow 2Fe^{3+}(aq) + 6I^-(aq)$
A and **C**: $3H_2O_2(aq) + 6H^+(aq) + 2Fe(s) \longrightarrow 2Fe^{3+}(aq) + 6H_2O(l)$

2 a) $Ni(s)$, $Fe^{2+}(aq)$, $Br^-(l)$, $H_2O(l)$.
b) Br_2 with Fe^{2+}; Fe^{3+} with Ni; O_2 and H^+ with Fe^{2+}; Br_2 with Ni; O_2 and H^+ with Br^-; O_2 and H^+ with Ni.

Redox processes in the nitrogen cycle

SALTERS MS

Chemists benefit agriculture by developing chemicals to:

- provide extra nutrients (fertilisers)
- control soil pH
- control pests (insecticides and pesticides).

In the nitrogen cycle, nitrogen is present in many forms and oxidation states, e.g.

gases:			aqueous ions:		
nitrogen	N_2	0	ammonium	NH_4^+	−3
ammonia	NH_3	−3	nitrate(III)	NO_2^-	+3
nitrogen(I) oxide	N_2O	+1	nitrate(V)	NO_3^-	+5
nitrogen(II) oxide	NO	+2			
nitrogen(IV) oxide	NO_2	+4			

In the environment, many redox processes involve nitrogen.

Fixation and nitrification

Conversion of nitrogen gas into soluble nitrogen-containing ions:

N_2	→	NH_4^+	→	NO_2^-	→	NO_3^-
0		−3		+3		+5

Denitrification

Conversion of soluble nitrogen-containing ions into nitrogen gas:

NO_3^-	→	NO_2^-	→	NO	→	N_2O	→	N_2
+5		+3		+2		+1		0

Advantages and disadvantages of fertilisers

Advantages

The soluble nitrogen content of the soil can be enhanced by adding 'artificial' fertilisers containing ammonium and nitrate ions.

- Extra nitrogen nutrients are provided, improving food production.
- Large quantities of fertilisers are produced by the chemical industry.
- Natural fertilisers, based on manure, or crop rotation, can improve soil structure.

Disadvantages

Nitrate fertilisers are soluble and can be leached out of the soil.

- Excessive use of artificial fertilisers can reduce the quality of soil structure.
- Nitrates can run off fields into rivers and ponds – a source of water pollution. Strategies, such as preventing the run off, can be put into place to reduce and manage water pollution.
- Excess nitrogen in watercourses can lead to eutrophication – excessive plant growth and decay causing depletion of dissolved oxygen needed for the survival of other aquatic life, such as fish.

Sample question and model answer

The standard electrode potentials of three half reactions are given below:

$MnO_4^-(aq) + 8H^+(aq) + 5e^- \rightleftharpoons Mn^{2+}(aq) + 4H_2O(aq)$ $E^\ominus = +1.52$ V

$Cl_2(g) + 2e^- \rightleftharpoons 2Cl^-(aq)$ $E^\ominus = +1.36$ V

$Fe^{3+}(aq) + e^- \rightleftharpoons Fe^{2+}(aq)$ $E^\ominus = +0.77$ V

To form Cl_2:
$Cl_2 + 2e^- \leftarrow 2Cl^-$
This reaction has been reversed and must be a negative electrode.
It can react with a half equation in the opposite direction with a more positive electrode potential:
$MnO_4^- + 8H^+ + 5e^- \rightarrow$

(a) $Cl_2(g)$ is produced in a redox process.

(i) Justify whether the conversion of $Cl^-(aq)$ to $Cl_2(g)$ is oxidation or reduction.

Oxidation because electrons are lost ✓ during the conversion.

(ii) Suggest what ions could be used for this conversion.

$MnO_4^-(aq)/H^+(aq)$ ✓

(iii) State the conditions needed.

acidified ✓

(iv) What chemicals could be used to produce $Cl_2(g)$?

$KMnO_4$ ✓ and concentrated HCl ✓ [5]

'Chemicals' or 'reagents' refer to actual materials that you would use in the laboratory to carry out a reaction.
Here, $KMnO_4$ and HCl are the reagents although only MnO_4^- and H^+ are involved in the reaction

(b) (i) Calculate the standard cell potential for the reaction between $Fe^{2+}(aq)$ and acidified $MnO_4^-(aq)$.

$E^\ominus_{cell} = E^\ominus(+ \text{ terminal}) - E^\ominus(- \text{ terminal})$

$\therefore E^\ominus_{cell} = +1.52 - (+0.77) = 0.75$ V ✓

Simply subtract the $E^\ominus$ of the negative terminal from the $E^\ominus$ of the positive terminal.

(ii) Construct a balanced equation for this reaction.

$MnO_4^-(aq) + 8H^+(aq) + 5e^- \rightarrow Mn^{2+}(aq) + 4H_2O(l)$

$Fe^{2+}(aq) \rightarrow Fe^{3+}(aq) + e^-$

Balance electrons: $5Fe^{2+}(aq) \rightarrow 5Fe^{3+}(aq) + 5e^-$

$5Fe^{2+}(aq) + MnO_4^-(aq) + 8H^+(aq) \rightarrow 5Fe^{3+}(aq) + Mn^{2+}(aq) + 4H_2O(l)$

use of (x 5) to balance electrons ✓ balanced equation ✓ [3]

Reverse the more negative half-reaction to give each half-cell reactions.

The overall cell reaction can be found by adding the balanced half equations.

(c) Suggest why hydrochloric acid is not used to acidify the $MnO_4^-(aq)$ in the reaction in (b)(i).

MnO_4^- would react with the Cl^- ✓ present in hydrochloric acid. [1]

[Total: 9]

Practice examination questions

OCR

1 The Born–Haber cycle for the formation of potassium bromide from potassium and bromine may be represented by a series of stages labelled **A** to **F** as shown.

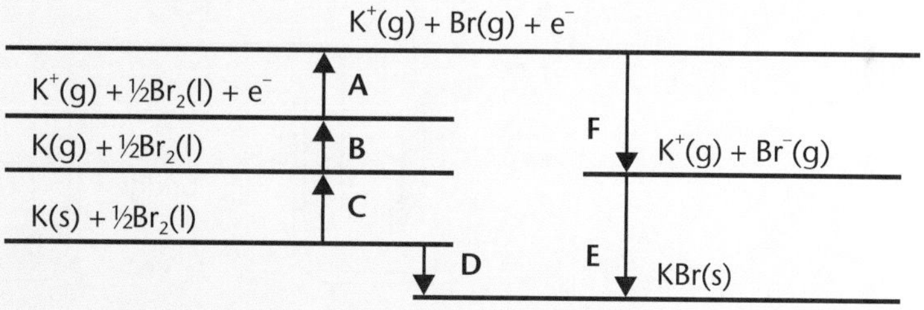

(a) (i) Write the letters **A** to **F** next to the corresponding definition in the table below.

definition	*letter*	*ΔH/kJ mol^{-1}*
1st ionisation energy of potassium		+419
1st electron affinity of bromine		–325
the enthalpy of atomisation of potassium		+89
the enthalpy of atomisation of bromine		+112
the lattice enthalpy of potassium bromide		
the enthalpy of formation of potassium bromide		–394

(ii) Calculate the lattice enthalpy of potassium bromide from the data given. [5]

(b) Explain whether you would expect the lattice enthalpies of the following compounds to be more or less exothermic than that of potassium bromide.

(i) KI; (ii) $CaBr_2$ [4]

[Total: 9]

2 (a) Write equations, including state symbols, for the following:

(i) the lattice enthalpy of magnesium chloride

(ii) the enthalpy change of hydration of a Mg^{2+} ion

(iii) the enthalpy change of solution of magnesium chloride. [3]

(b) The lattice enthalpy of $MgCl_2$ is –2526 kJ mol^{-1}.

The enthalpy change of hydration of Mg^{2+} is –1891 kJ mol–1.

The enthalpy change of hydration of Cl^- is –384 kJ mol–1.

Calculate the enthalpy change of solution of magnesium chloride. [3]

[Total: 6]

3 Chromium can be prepared by the reduction of chromium(III) oxide with aluminium.

$$Cr_2O_3(s) + 2Al(s) \longrightarrow 2Cr(s) + Al_2O_3(s)$$

The table below gives information about the reactants and products.

substance	*$Cr_2O_3(s)$*	*Al(s)*	*Cr(s)*	*$Al_2O_3(s)$*
ΔH_f/kJ mol^{-1}	–602	0	0	–1676
S/J K^{-1} mol^{-1}	27	28	81	51

Calculate the free energy change for this reaction at 298 K and comment on its feasibility at this temperature.

[Total: 8]

Practice examination questions (continued)

4 A standard cell was set up as shown below.

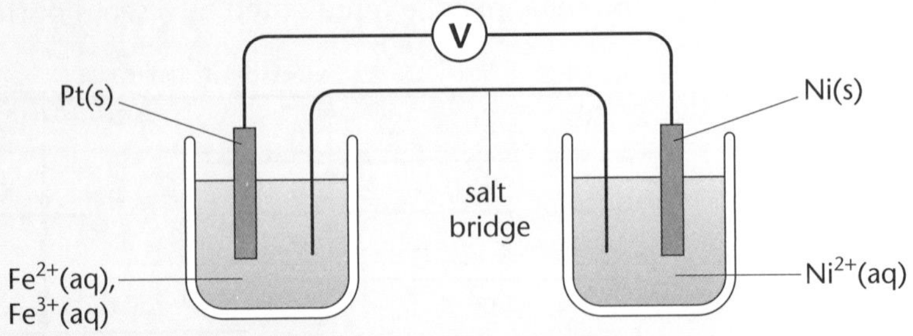

(a) What conditions are necessary for this cell to be a standard cell? [2]

(b) What is the purpose of the salt bridge? [1]

(c) The standard electrode potentials for the half-cells above are:

$Ni^{2+} + 2e^- \rightleftharpoons Ni$ $E^\ominus = -0.25$ V

$Fe^{3+} + e^- \rightleftharpoons Fe^{2+}$ $E^\ominus = +0.77$ V

(i) Calculate the standard cell potential of the complete cell.

(ii) Draw an arrow on the diagram above to show the direction of electron flow in the external circuit.

(iii) What is the polarity of each electrode? [3]

(d) (i) Write equations for the reaction at each electrode.

(ii) Hence write an equation to show the overall reaction. [3]

(e) The solution in the nickel half-cell was replaced by a solution that had a higher concentration of Ni^{2+}(aq) ions. Predict the effect on the cell potential. Explain your reasoning. [2]

[Total: 11]

5 (a) Use the standard electrode potentials given below to answer the questions that follow.

			$E^\ominus$/ V
$Mg^{2+}(aq) + 2e^-$	$\rightleftharpoons$	Mg(s)	–2.37
$I_2(aq) + 2e^-$	$\rightleftharpoons$	$2I^-(aq)$	+0.54
$Mn^{3+}(aq) + e^-$	$\rightleftharpoons$	$Mn^{2+}(aq)$	+1.49

(i) Which of the six species above is the strongest oxidising agent?

(ii) Predict, with an equation and a reason, which species above would react with I^-(aq) ions to form iodine, I_2(aq). [4]

(b) The standard electrode potentials for two redox equilibria are given below.

$O_2(g) + 2H_2O(l) + 4e^- \rightleftharpoons 4OH^-(aq)$ $E^\ominus = +0.40$ V

$V^{3+}(aq) + e^- \rightleftharpoons V^{2+}(aq)$ $E^\ominus = -0.26$ V

Two half-cells based on these redox equilibria are connected to make a cell. Determine the cell potential, $E^\ominus$, for this cell and write an equation for the cell reaction. [3]

[Total: 7]

Chapter 3

The Periodic Table

The following topics are covered in this chapter:

- *The transition elements*
- *Transition element complexes*
- *Ligand substitution of complex ions*
- *Redox reactions of transition metal ions*
- *Catalysis*
- *Reactions of metal aqua-ions*

3.1 The transition elements

After studying this section you should be able to:

- *explain what is meant by the terms d-block element and transition element*
- *deduce the electronic configurations of atoms and ions of the d-block elements Sc → Zn*
- *know the typical properties of a transition element*

LEARNING SUMMARY

Key points from AS

- **The Periodic Table** *Revise AS page 61–70*

During AS Chemistry, you studied the metals in the s-block of the Periodic Table. For A2 Chemistry, this understanding is extended with a detailed study of the elements scandium to zinc in the d-block of the Periodic Table. Many of the principles introduced during AS Chemistry will be revisited, especially redox chemistry, and structure and bonding.

The d-block elements

OCR M5
SALTERS M4

The d-block is found in the centre of the Periodic Table. For A2 Chemistry, you will be studying mainly the d-block elements in Period 4 of the Periodic Table: Sc, Ti, V, Cr, Mn, Fe, Co, Ni, Cu and Zn.

Electronic configurations of d-block elements

The diagram below shows the sub-shell being filled across Period 4 of the Periodic Table.

4s		3d										4p					
K	Ca	Sc	Ti	V	Cr	Mn	Fe	Co	Ni	Cu	Zn	Ga	Ge	As	Se	Br	Kr

Key points from AS

- **Sub-shells and orbitals** *Revise AS pages 17–19*
- **Filling the sub-shells** *Revise AS pages 19–20*
- **Sub-shells and the Periodic Table** *Revise AS page 20*

The 3d sub-shell is at a higher energy and fills **after** the 4s sub-shell.

See also the stability from a half-full p sub-shell.

- The elements Sc ⟶ Zn have their highest energy electrons in the 4s and 3d sub-shells.
- Across the d-block, electrons are filling the 3d sub-shell. The diagram below shows the outermost electron shell for Sc ⟶ Zn.

Notice how the orbitals are filled:

- the orbitals in the 3d sub-shell are first occupied singly to prevent any repulsion caused by pairing.

Notice chromium and copper:

- chromium has one electron in each orbital of the 4s and 3d sub-shells ⟶ extra stability
- copper has a full 3d sub-shell ⟶ extra stability.

		4s	3d				
Sc	[Ar] $3d^1 4s^2$	↑↓	↑				
Ti	[Ar] $3d^2 4s^2$	↑↓	↑	↑			
V	[Ar] $3d^3 4s^2$	↑↓	↑	↑	↑		
Cr	[Ar] $3d^5 4s^1$	↑	↑	↑	↑	↑	↑
Mn	[Ar] $3d^5 4s^2$	↑↓	↑	↑	↑	↑	↑
Fe	[Ar] $3d^6 4s^2$	↑↓	↑↓	↑	↑	↑	↑
Co	[Ar] $3d^7 4s^2$	↑↓	↑↓	↑↓	↑	↑	↑
Ni	[Ar] $3d^8 4s^2$	↑↓	↑↓	↑↓	↑↓	↑	↑
Cu	[Ar] $3d^{10} 4s^1$	↑	↑↓	↑↓	↑↓	↑↓	↑↓
Zn	[Ar] $3d^{10} 4s^2$	↑↓	↑↓	↑↓	↑↓	↑↓	↑↓

Transition elements

Successive ionisation energies provide evidence for electronic configurations of the d-block elements Sc to Zn. This is most evident by the presence of peaks matching the $3d^5$ half-filled shell.

The elements in the d-block, Ti→Cu, are called transition elements.

> **KEY POINT**
> • A transition element has **at least one ion** with a **partially-filled d sub-shell.**

Although scandium and zinc are d-block elements, they are **not** *transition elements.* Neither element forms an ion with a partially filled d sub-shell.

Scandium forms one ion only, Sc^{3+} with the 3d sub-shell **empty.**

$$\text{Sc [Ar]}3d^14s^2 \xrightarrow{\textit{loss of 3 electrons}} Sc^{3+} \text{ only: [Ar]}$$

Zinc forms one ion only, Zn^{2+} with the 3d sub-shell **full.**

$$\text{Zn [Ar]}3d^{10}4s^2 \xrightarrow{\textit{loss of 2 electrons}} Zn^{2+} \text{ only: [Ar]}3d^{10}$$

d–block elements

Sc	Ti	V	Cr	Mn	Fe	Co	Ni	Cu	Zn

transition elements (Ti to Cu)

Key points from AS

• **The s-block elements** *Revise AS pages 65–67*

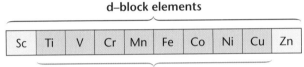

Typical properties of transition elements

Properties of transition elements

1 high density

2 high m. pt. and b. pt.

3 moderate to low reactivity

4 two or more oxidation states

5 coloured ions

6 complex ions with ligands

7 catalysts

The transition elements are all metals – they are good conductors of heat and electricity. In addition, some general properties distinguish the transition elements from the s-block metals.

Density

- The transition elements are more **dense** than other metals.
  - They have smaller atoms than the metals in Groups 1 and 2. The small atoms are able to pack closely together ⟶ high density.

Melting point and boiling point

- The transition elements have **higher melting and boiling points** than other metals.
  - Within the metallic lattice, ions are **smaller** than those of the s-block metals – the metallic bonding between the metal ions and the delocalised electrons is strong.

Reactivity

- The transition elements have **moderate to low reactivity.**
  - Unlike the s-block metals, they do **not** react with cold water. Many transition metals react with dilute acids although some, such as gold, silver and platinum, are extremely unreactive.

Oxidation states

- Most transition elements have compounds with **two or more oxidation states.**

Colour of compounds

- The transition elements have **coloured compounds and ions.**
  - Most transition metals have at least one oxidation state that is coloured.

Complex ions

- The ions of transition elements form **complex ions** with ligands.

Catalysis

- Many transition elements can act as **catalysts.**

Formation of ions

OCR M5
SALTERS M4

The energy levels of the 4s and 3d sub-shells are very close together.

KEY POINT

- Transition elements are able to form **more than one ion**, each with a different oxidation state, by losing the **4s** electrons and different numbers of **3d** electrons.
- When forming ions, the 4s electrons are **lost first**, before the 3d electrons.

The Fe^{3+} ion, $[Ar]3d^5$ is more stable than the Fe^{2+} ion, $[Ar]3d^5 4s^1$.
Fe^{3+} has a half-full 3d sub-shell ⟶ stability.

The 4s and 3d energy levels are very close together and both are involved in bonding.
When forming ions, the 4s electrons are **lost before** the 3d electrons.
Note that this is different from the order of **filling** – 4s before 3d.
Once the 3d sub-shell starts to fill, the 4s electrons are repelled to a higher energy level and are lost first.

Forming Fe^{2+} and Fe^{3+} ions from iron

Iron has the electronic configuration $[Ar]3d^6 4s^2$

Iron forms two common ions, Fe^{2+} and Fe^{3+}.

An Fe atom loses **two 4s electrons** ⟶ Fe^{2+} ion.

An Fe atom loses **two 4s electrons** and **one 3d electron** ⟶ Fe^{3+} ion.

The variety of oxidation states

The table below summarises the stable oxidation states of the elements scandium to zinc. Those in bold type represent the commonest oxidation numbers of the elements in their compounds.

Sc	Ti	V	Cr	Mn	Fe	Co	Ni	Cu	Zn
	+1	+1	+1	+1	+1	+1	+1	+1	
	+2	+2	+2	+2	+2	+2	+2	+2	+2
+3	+3	+3	+3	+3	+3	+3	+3	+3	
	+4	+4	+4	+4	+4	+4	+4		
		+5	+5	+5	+5	+5			
			+6	+6	+6				
				+7					

- The variety of oxidation states results from the close similarity in energy of the 4s and the 3d electrons.
- The number of available oxidation states for the element **increases** from Sc → Mn.
  - All of the available 3d and 4s electrons may be used for bond formation.
- The number of available oxidation states for the element **decreases** from Mn → Zn.
  - There is a decreasing number of unpaired d-electrons available for bond formation.
- Higher oxidation states involve covalency because of the high charge densities involved.
  - In practice, for oxidation states of +4 and above, the bonding electrons are involved in covalent bond formation.

Refer to the electronic configurations of the d-block elements, shown on page 67.

Progress check

1 Write down the electronic configuration of the following ions:
(a) Ni^{2+} (b) Cr^{3+} (c) Cu^{+} (d) V^{3+}.

2 What is the oxidation state of the metal in each of the following?
(a) MnO_2 (b) CrO_3 (c) MnO_4^- (d) VO_2^+

1 (a) $[Ar]3d^8$ (b) $[Ar]3d^3$ (c) $[Ar]3d^{10}$ (d) $[Ar]3d^2$.
2 (a) +4 (b) +6 (c) +7 (d) +5.

3.2 Transition element complexes

After studying this section you should be able to:

LEARNING SUMMARY

- *explain what is meant by the terms 'complex ion' and 'ligand'*
- *predict the formula and possible shape of a complex ion*
- *know that ligands can be unidentate, bidentate and multidentate*
- *understand that some complexes have stereoisomers*
- *know that transition metal ions can be identified by their colour*
- *know that electronic transitions are responsible for colour*
- *explain how colorimetry can be used to determine the concentration and formula of a complex ion*

Ligands and complex ions

OCR M5
SALTERS M4

Transition metal ions are small and densely charged. They strongly attract electron-rich species called **ligands** forming **complex ions**.

A ligand has a lone pair of electrons.

KEY POINT

A ligand is a molecule or ion that bonds to a metal ion by:

- forming a coordinate (dative covalent) bond
- donating a lone pair of electrons into a vacant orbital.

Common ligands include: H_2O:, :Cl^-, :NH_3, :CN^-

KEY POINT

A **complex ion** is a central metal ion surrounded by ligands.
The **coordination number** is the total **number** of **coordinate bonds** from ligands to the central transition metal ion of a complex ion.

Key points from AS

- **Electron pair repulsion theory**
  Revise AS pages 50–51

For AS, you learnt that electron pair repulsion is responsible for the shape of a molecule or ion.

Shapes of complex ions

The size of a ligand helps to decide the shape of the complex ion.

Six coordinate complexes

Six water molecules are able to fit around a Cu^{2+} ion to form:

- the complex ion $[Cu(H_2O)_6]^{2+}$
- with a coordination number of **6**.

The **six** electron pairs surrounding the central Cu^{2+} ion in $[Cu(H_2O)_6]^{2+}$ repel one another as far apart as possible to form a complex ion with an **octahedral** shape.

The hexaaquacopper(II) complex ion, $[Cu(H_2O)_6]^{2+}$

$[Cu(H_2O)_6]^{2+}$
octahedral
6 coordinate

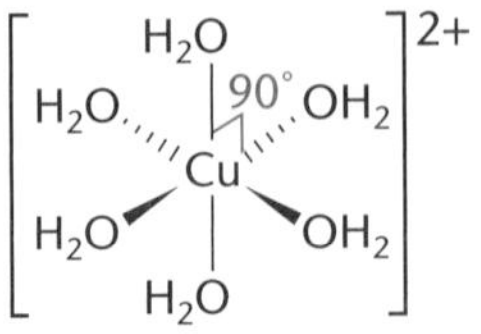

Four coordinate complexes

Chloride ions are larger than water molecules and it is only possible for **four** chloride ions to fit around the central copper(II) ion to form:

- the complex ion $[CuCl_4]^{2-}$
- with a coordination number of **4**.

The **four** electron pairs surrounding the central Cu^{2+} ion in $[CuCl_4]^{2-}$ repel one another as far apart as possible to form a complex ion with a **tetrahedral** shape.

The tetrachlorocuprate(II) complex ion $[CuCl_4]^{2-}$

Notice the overall charge on the complex ion $[CuCl_4]^{2-}$ is 2–.

Cu^{2+} and 4 Cl^- → 2– ions.

$[CuCl_4]^{2-}$
tetrahedral
4 coordinate

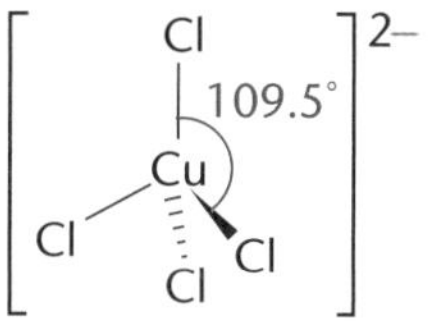

General rules for deciding the shape of a complex ion

Although there are exceptions, the following general rules are useful.

> **KEY POINT**
>
> Complex ions with small ligands such as H_2O and NH_3 are usually **6-coordinate** and **octahedral**.
>
> Complex ions with large Cl^- ligands are usually **4-coordinate** and **tetrahedral**.

The table below compares complex ions formed from cobalt(II) and chromium(III) ions.

ligand	*complex ions from Co^{2+}*	*complex ions from Cr^{3+}*	*shape*
H_2O	$[Co(H_2O)_6]^{2+}$	$[Cr(H_2O)_6]^{3+}$	octahedral
NH_3	$[Co(NH_3)_6]^{2+}$	$[Cr(NH_3)_6]^{3+}$	octahedral
Cl^-	$[CoCl_4]^{2-}$	$[CrCl_4]^-$	tetrahedral

- Some complex ions contain more than one type of ligand. For example, copper(II) forms a complex ion with a mixture of water and ammonia ligands, $[Cu(NH_3)_4(H_2O)_2]^{2+}$.

Ligands have teeth

OCR M5
SALTERS M4

Ligands such as H_2O, NH_3 and Cl^- are called **monodentate** ligands. They form only **one** coordinate bond to the central metal ion.

A molecule or ion with more than one oxygen or nitrogen atom may form more than one coordinate bond to the central metal ion.

Ligands that can form **two** coordinate bonds to the central metal ion are called **bidentate** ligands. Examples of bidentate ligands are ethane-1,2-diamine, $NH_2CH_2CH_2NH_2$ and the ethanedioate ion, $(COO^-)_2$.

Monodentate means **one** tooth.

Bidentate ligands have **two** teeth.

Multidentate ligands have **many** teeth.

Each 'tooth' is a coordinate bond.

Multidentate or polydentate ligands can form **many** coordinate bonds to the central metal ion. For example, the **hexa**dentate ligand, edta^{4-}, is able to form **six** coordinate bonds to the central metal ion. The diagram of edta^{4-} shows four oxygen atoms and two nitrogen atoms able to form coordinate bonds.

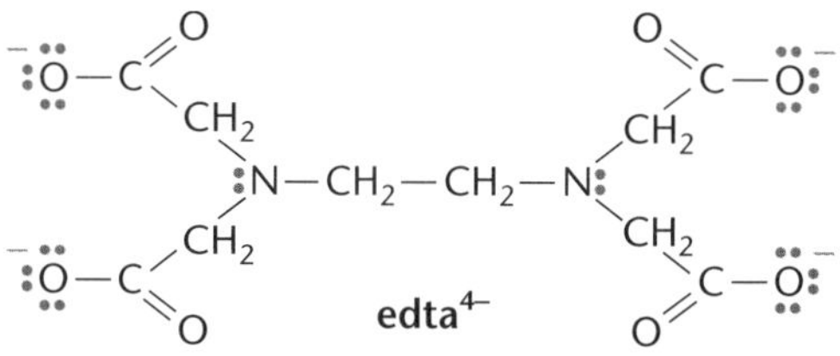

edta^{4-}

Stereoisomerism in complexes

OCR M5

Stereoisomers have the same structural isomers but different arrangements in space. Transition metal complexes show two types of stereoisomerism:

- *cis-trans* isomerism
- optical isomerism.

Cis-trans isomerism

Cis-trans isomerism occurs in square planar complexes with two ligands of one kind and two of another kind, e.g. $Ni(NH_3)_2Cl_2$.

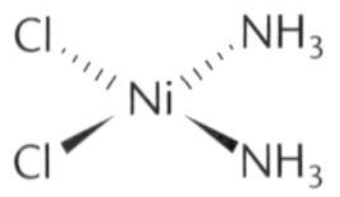

cis-isomer
groups on same side

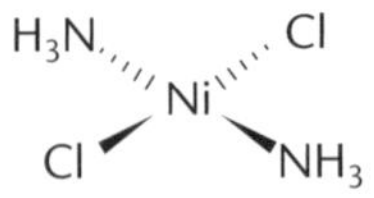

trans-isomer
groups on opposite side

Platin

The structures of *cis*-platin and *trans*-platin are shown below. Both structures have a square planar shape with a 90° bond angle.

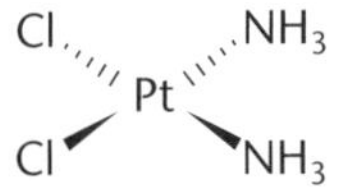

cis-platin

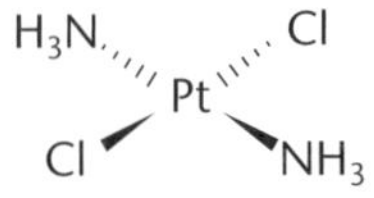

trans-platin

Cis-platin is used as a chemotherapy drug in the treatment of some cancers. *Cis*-platin forms coordinate bonds to a basic site in the DNA of cancer cells. This process prevents the cancerous DNA from replicating, effectively killing the cancerous cells. Careful treatment is required to avoid harmful side-effects of the drug.

Optical isomerism

Optical isomers are non-superimposable mirror images of one another. Some octahedral complexes with multidentate ligands form optical isomers, e.g. $[Ni(NH_2CH_2CH_2NH_2)_3]^{2+}$.

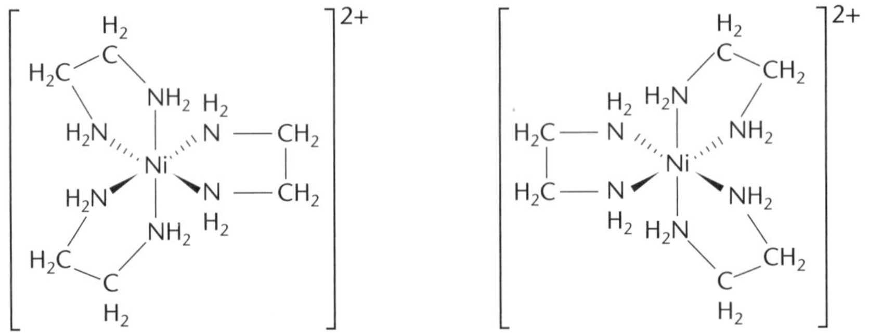

The two optical isomers rotate plane-polarised light in opposite directions.

Colour and complex ions

OCR M4
SALTERS M5

Complex ions have different colours

Many transition metal ions have different colours. These colours indicate the possible identity of a transition metal ion. The table on page 73 shows the colours of some **aqua** complex ions of d-block elements. Notice that different oxidation states of a d-block element can have different colours.

Sc	Ti	V	Cr	Mn	Fe	Co	Ni	Cu	Zn
								+1 *colourless*	
				+2 *pale pink*	+2 *pale green*	+2 *pink*	+2 *green*	+2 *blue*	+2 *colourless*
+3 *colourless*	+3 *violet*	+3 *green*	+3 **ruby/ green*		+3 *red-brown*				
		+5 yellow							
			+6 *yellow or orange*						
				+7 *purple*					

* Cr^{3+} surrounded by 6 water ligands looks violet, but some water is often replaced by other ligands so Cr^{3+} usually looks green.

Why are transition element ions coloured?

SALTERS MS

Splitting the 3d energy level

All five d-orbitals in the outer d sub-shell of an **isolated** transition element ion have the same energy. However, when ligands bond to the central ion, the d-orbitals move to two different energy levels. The gap between the energy levels depends upon:

- the ligand
- the coordination number
- the transition metal ion.

This diagram shows how water ligands split the energy levels of the five 3d orbitals in a copper(II) ion.

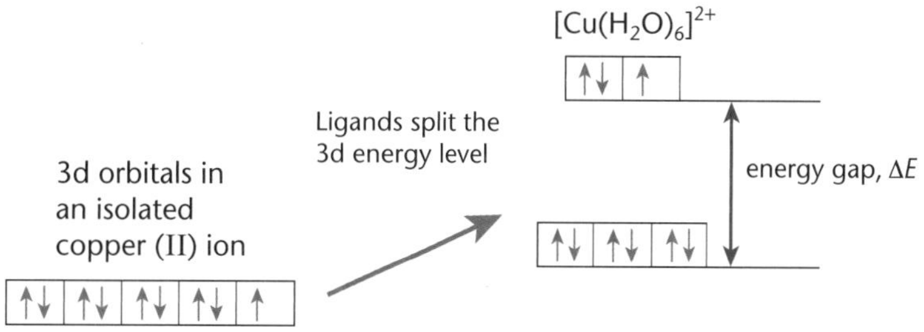

Absorption of light energy

An electron can be promoted from the lower 3d energy level by absorbing energy **exactly** equal to the energy gap ΔE. This energy is provided by radiation from the visible and ultraviolet regions of the spectrum.

Radiation with a frequency f has an energy hf, where h = Planck's constant, 6.63×10^{-34} J s.

So, to promote an electron through as energy gap ΔE, a quantity of light energy is needed given by:

$$\Delta E = hf$$

Energy and frequency

The difference in energy between the d-orbitals, ΔE, and the frequency of the absorbed radiation, f, are linked by the following relationship:

$\Delta E = hf$

h = Planck's constant, 6.63×10^{-34} J s.

The diagram below shows what happens when a complex ion absorbs light energy.

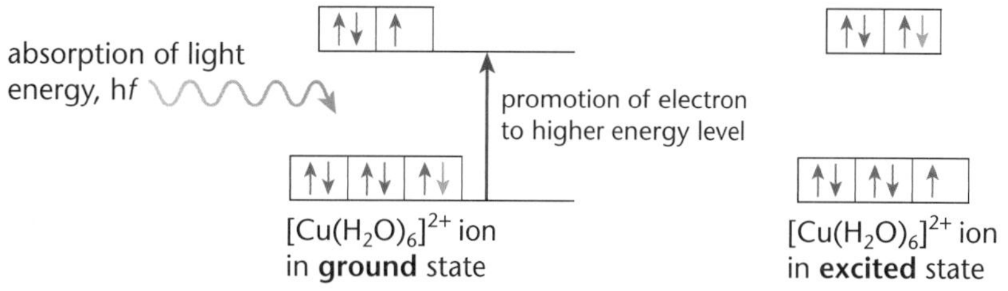

Only those ions that have a partially filled d sub-shell are coloured.
Cu^{2+} is $[Ar]3d^94s^2$

The blue colour of $[Cu(H_2O)_6]^{2+}$ results from:

- **absorption** of light energy in the red, yellow and green regions of the spectrum. This absorbed radiation provides the energy for an electron to be excited to a higher energy level
- **reflection** of blue light only, giving the copper(II) hexaaqua ion its characteristic blue colour. The blue region of the spectrum is **not** absorbed.

Copper has another ion, Cu^+, with the electronic configuration: $[Ar]3d^{10}$.

The Cu^+ ion is colourless because:

- the 3d sub-shell is full, $[Ar]3d^{10}$, preventing electron transfer.

Ag^+ complexes are also colourless because the 3d sub-shell is full.

> The colour of a transition metal complex ion results from the **transfer** of an **electron** between the orbitals of an **unfilled** d sub-shell.
>
> KEY POINT

Absorbance and reflectance spectra

The absorbance and reflectance spectra of a pigment are shown below. Note that the spectra show peaks the opposite way around.

Key points from AS

- **Atomic spectra of hydrogen**
  Revise AS pages 23–24

Different techniques can be used to piece together analytical information about substances.

For example, analysis of paintings may include:

- gas chromatography, GC (see pages 141–142)
- atomic emission and absorption spectroscopy (see AS pages 23–24)
- visible spectroscopy (reflection and absorption).

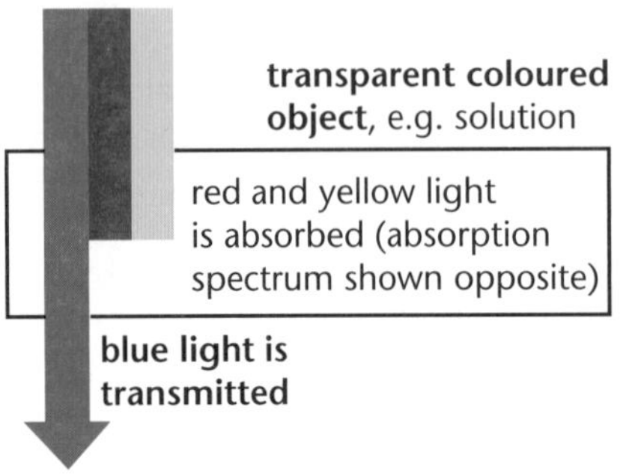

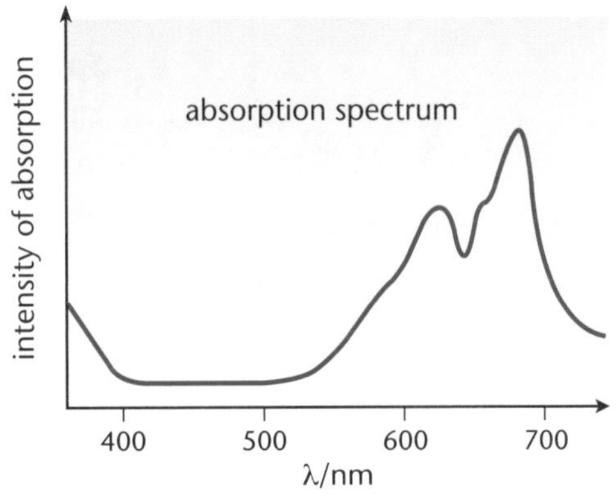

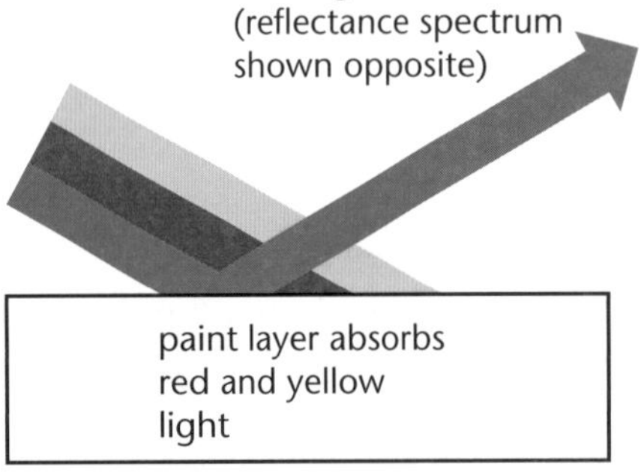

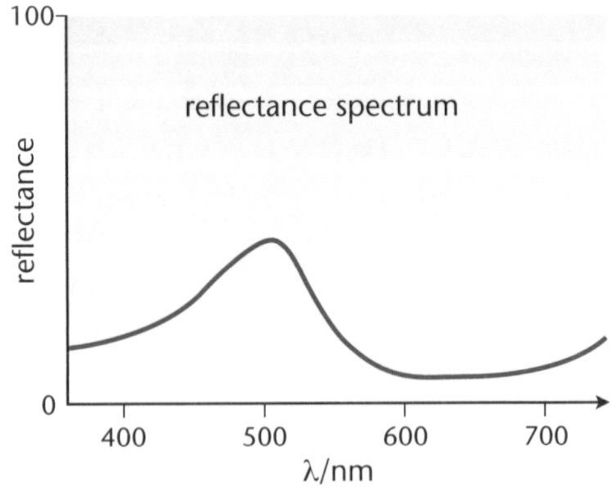

Using colorimetry

SALTERS M4

A colorimeter measures the amount of light absorbed by a coloured solution. The concentration and formula of a complex ion can be determined from the intensity of light absorbed by the colorimeter.

Finding an unknown concentration of a coloured solution

- A suitable filter and wavelength is chosen for the readings.
- Solutions of a coloured complex ion of known concentration are prepared.

- The absorbances of the solutions of known concentration are measured.
- A calibration curve is plotted.
- The solution of the complex ion with an unknown concentration is then placed in the colorimeter and the absorbance measured.
- The unknown concentration can be determined from the calibration curve.

Colorimetry can also be used as a method for monitoring reaction rates (see pages 14–23).

Progress check

1 Predict the formula of the following complex ions:
(a) iron(III) with water ligands
(b) vanadium(III) with chloride ligands
(c) nickel(II) with ammonia ligands.

SALTERS

2 Explain the origin of colour in a complex ion that is yellow.

1 (a) $[Fe(H_2O)_6]^{3+}$ (b) $[VCl_4]^-$ (c) $[Ni(NH_3)_6]^{2+}$.

2 The 3d orbitals are split between two energy levels. A 3d electron from the lower 3d energy level absorbs visible light in the blue and red regions of the spectrum. The absorbed energy allows the d-electron to be promoted to a vacancy in the orbitals of the higher 3d energy level. With the red and blue regions of the spectrum absorbed, only yellow light is reflected, giving the complex ion its colour.

3.3 Ligand substitution of complex ions

After studying this section you should be able to:

- *describe what is meant by ligand substitution*
- *understand that ligand exchange may produce changes in colour and coordination number*
- *understand stability of complex ions in terms of stability constants*

LEARNING SUMMARY

Exchange between ligands

OCR M5
SALTERS M4

A **ligand substitution** reaction takes place when a ligand in a complex ion exchanges for another ligand. A change in ligand usually changes the energy gap between the 3d energy levels. With a different ΔE value, light with a different frequency is absorbed, producing a colour change.

Ligand substitution usually produces a change in colour.

Exchange between H_2O and NH_3 ligands

The **similar sizes** of water and ammonia molecules ensure that ligand exchange takes place with **no change in coordination number**.

For example, addition of an excess of concentrated aqueous ammonia to aqueous nickel(II) ions results in ligand substitution of ammonia ligands for water ligands. Both complex ions have an octahedral shape with a coordination number of six.

Remember that the **size** of the ligand is a major factor in deciding the coordination number (see page 70):

- H_2O and NH_3 have similar sizes – same coordination number
- H_2O and Cl^- have different sizes – different coordination numbers.

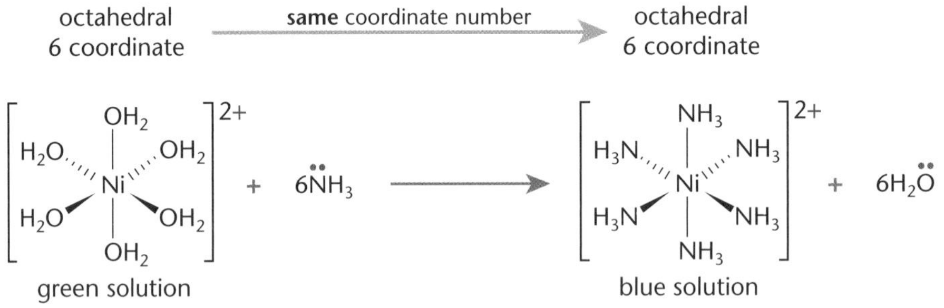

$[Ni(H_2O)_6]^{2+} + 6NH_3 \longrightarrow [Ni(NH_3)_6]^{2+} + 6H_2O$

Exchange between H_2O and Cl^- ligands

The **different sizes** of water molecules and chloride ions ensure that ligand exchange takes place **with a change in coordination number**.

For example, addition of an excess of concentrated hydrochloric acid (as a source of Cl^- ions) to aqueous cobalt(II) ions results in ligand substitution of chloride ligands for water ligands. The complex ions have different shapes with different coordination numbers.

You should also be able to construct a balanced equation for any ligand exchange reaction.

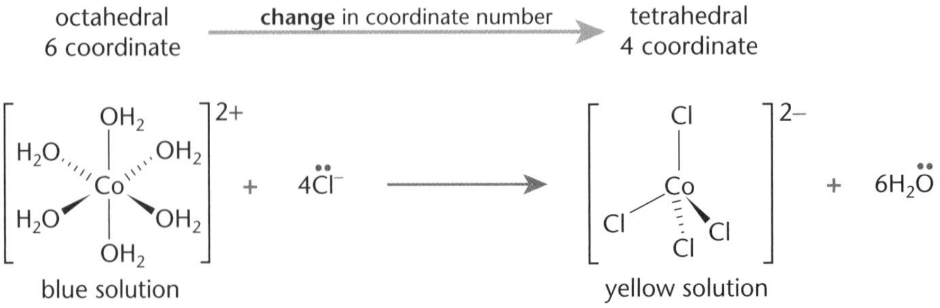

$[Co(H_2O)_6]^{2+} + 4Cl^- \longrightarrow [CoCl_4]^{2-} + 6H_2O$

Ligand substitutions of copper(II) and cobalt(II)

Concentrated hydrochloric acid is used as a source of Cl^- ligands.
Concentrated aqueous ammonia is used as a source of NH_3 ligands.

Ligand substitutions of copper(II) and cobalt(II) complexes are shown below.

Copper

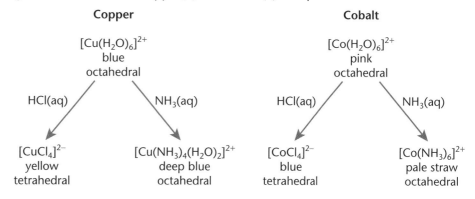

$[Cu(H_2O)_6]^{2+}$ blue octahedral

- $HCl(aq)$ → $[CuCl_4]^{2-}$ yellow tetrahedral
- $NH_3(aq)$ → $[Cu(NH_3)_4(H_2O)_2]^{2+}$ deep blue octahedral

Cobalt

$[Co(H_2O)_6]^{2+}$ pink octahedral

- $HCl(aq)$ → $[CoCl_4]^{2-}$ blue tetrahedral
- $NH_3(aq)$ → $[Co(NH_3)_6]^{2+}$ pale straw octahedral

Incomplete substitution

This reaction is used as a test for the Fe^{3+} ion – the deep red-brown colour of $[Fe(H_2O)_5(SCN)]^{2+}$ is intense and extremely small traces of Fe^{3+} ions can be detected using this ligand substitution reaction.

Substitution may be incomplete.

- Aqueous ammonia only exchanges **four** from the six water ligands in $[Cu(H_2O)_6]^{2+}$.

$$[Cu(H_2O)_6]^{2+} + 4NH_3 \longrightarrow [Cu(NH_3)_4(H_2O)_2]^{2+} + 4H_2O$$

blue solution → deep blue solution

- Aqueous thiocyanate ions, SCN^-, only exchanges **one** from the six water ligands in $[Fe(H_2O)_6]^{3+}$.

$$[Fe(H_2O)_6]^{3+} + SCN^-(aq) \longrightarrow [Fe(H_2O)_5(SCN)]^{2+} + H_2O$$

pale brown solution → deep red solution

Haemoglobin

OCR MS

Haemoglobin is a complex ion of Fe^{2+}. A haem group consists of an Fe^{2+} ion, held in a ring, known as a porphyrin. The Fe^{2+} ion, which is the site of oxygen binding, bonds with four nitrogen atoms in the centre of the ring in a square planar geometry. The protein globin bonds below and O_2 is able to bond weakly above.

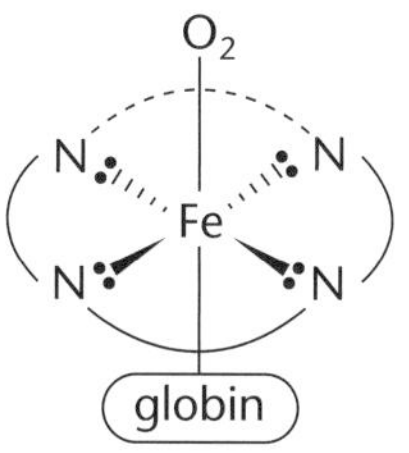

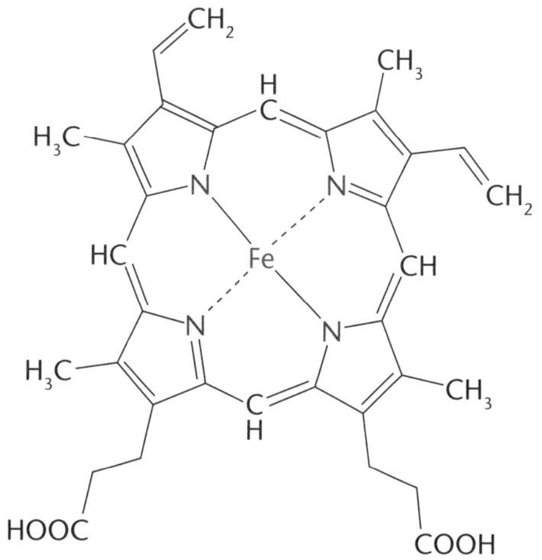

The porphyrin ring in haemoglobin

O_2 can be easily lost when required and is replaced by a weakly bonded water ligand. When haemoglobin reaches the lungs again, fresh O_2 is able to bond to Fe^{2+}, displacing the water.

Carbon monoxide, CO, is a serious poison for the blood as the polar molecule binds much more strongly to the Fe^{2+} ion than O_2 and cannot be displaced. This prevents oxygen from being carried by the blood.

Stability of complex ions

OCR MS

The stability of a complex ion depends upon the ligands. For example, a complex ion of copper(II) is more stable with ligands of ammonia than with water ligands.

Bidentate or multidentate ligands (such as edta) lead to a particularly stable complex. This is due to an increase in entropy:

$$\text{e.g.}\quad \underbrace{[Cu(H_2O)_6]^{2+} + edta^{4-}}_{\text{2 mol}} \xrightarrow[\text{ENTROPY}]{\text{INCREASE IN}} \underbrace{[Cu(edta)]^{2-} + 6H_2O}_{\text{7 mol}}$$

The concentration of water is virtually constant. Constant terms are not included in expressions for equilibrium constants.

The stability of a complex ion is measured as a **stability constant**, K_{stab}:

$$[Cu(H_2O)_6(aq)]^{2+} + 4NH_3(aq) \rightleftharpoons [Cu(NH_3)_4(H_2O)_2(aq)]^{2+} + 4H_2O(l)$$

$$K_{stab} = \frac{[[Cu(NH_3)_4(H_2O)_2(aq)]^{2+}]}{[[Cu(H_2O)_6(aq)]^{2+}\,[NH_3(aq)]^4]} = 1 \times 10^{12}\ dm^{12}\ mol^{-4}\ \text{at } 25°C$$

The data is expressed on a logarithmic scale because of the very large range of values.

Stability constants usually measure the stability of a complex compared with the metal aqua-ion. The greater the value of K_{stab}, the more stable the complex ion.

complex ion	*log K_{stab}*	
$[Fe(H_2O)_5(SCN)]^{2+}$	3.9	
$[CuCl_4]^{2-}$	5.6	increased
$[Ni(NH_3)_6]^{2+}$	8.6	stability
$[Ni(edta)]^{2-}$	19.3	

Progress check

1 Write equations for the following ligand substitutions:
(a) $[Ni(H_2O)_6]^{2+}$ with six ammonia ligands
(b) $[Fe(H_2O)_6]^{3+}$ with four chloride ligands
(c) $[Cr(H_2O)_6]^{3+}$ with six cyanide ligands, CN^-.

1 (a) $[Ni(H_2O)_6]^{2+} + 6NH_3 \longrightarrow [Ni(NH_3)_6]^{2+} + 6H_2O$
(b) $[Fe(H_2O)_6]^{3+} + 4Cl^- \longrightarrow [FeCl_4]^- + 6H_2O$
(c) $[Cr(H_2O)_6]^{3+} + 6CN^- \longrightarrow [Cr(CN)_6]^{3-} + 6H_2O$.

3.4 Redox reactions of transition metal ions

After studying this section you should be able to:

- *describe redox behaviour in transition elements*
- *construct redox equations, using relevant half-equations*
- *perform calculations involving simple redox titrations*

LEARNING SUMMARY

Redox reactions

OCR M5
SALTERS M4

Transition elements have variable oxidation states with characteristic colours (see page 73). Many **redox** reactions take place in which transition metals ions change their oxidation state by **gaining** or **losing** electrons.

The iron(II) – manganate(VII) reaction

Key points from AS

- **Redox reactions** *Revise AS pages 42–44*

Iron(II) ions are oxidised by acidified manganate(VII) ions.

- Acidified manganate(VII) ions are reduced to manganate(II) ions.

reduction: $MnO_4^-(aq) + 8H^+(aq) + 5e^- \longrightarrow Mn^{2+}(aq) + 4H_2O(l)$

+7 → +2

purple → very pale pink

The reaction is easy to see.

The deep purple MnO_4^- ions are reduced to the very pale pink Mn^{2+} ions, which are virtually colourless in solution.

- Iron(II) ions are oxidised to iron(III) ions.

oxidation: $Fe^{2+}(aq) \longrightarrow Fe^{3+}(aq) + e^-$

+2 → +3

To give the overall equation:

- the electrons are balanced

$$MnO_4^-(aq) + 8H^+(aq) + 5e^- \longrightarrow Mn^{2+}(aq) + 4H_2O(l)$$
$$5Fe^{2+}(aq) \longrightarrow 5Fe^{3+}(aq) + 5e^-$$

- the half-equations are added

$$5Fe^{2+}(aq) + MnO_4^-(aq) + 8H^+(aq) \longrightarrow 5Fe^{3+}(aq) + Mn^{2+}(aq) + 4H_2O(l)$$

Redox titrations

OCR M5
SALTERS M4

Redox titrations can be used in analysis.

Essential requirements are:

- an oxidising agent
- a reducing agent
- an easily-seen colour change (with or without an indicator).

Redox titrations using the acidified manganate(VII)

It is important to use an acid that does **not** react with either the oxidising or reducing agent – dilute sulphuric acid is usually used.

For example, hydrochloric acid cannot be used because HCl is oxidised to Cl_2 by MnO_4^-.

- $KMnO_4$ is reacted with a reducing agent such as Fe^{2+} under acidic conditions (see above).
- Purple $KMnO_4$ is added from the burette to a measured amount of the acidified reducing agent.
- At the end-point, the colour changes from colourless to pale pink, showing that all the reducing agent has exactly reacted. The pale-pink colour indicates the first trace of an excess of purple manganate(VII) ions.
- This titration is **self-indicating** – no indicator is required, the colour change from the reduction of MnO_4^- ions being sufficient to indicate that the reaction is complete.

The most common redox titration encountered at A Level is that between manganate(VII) and acidified iron(II) ions.

Key points from AS

- **Calculations in acid–base titrations**
  Revise AS page 34

Calculations for redox titrations follow the same principles as those used for acid–base titrations, studied during AS Chemistry.

The estimation of iron in iron tablets

Five iron tablets with a combined mass of 0.900 g were dissolved in acid and made up to 100 cm^3 of solution. In a titration, 10.0 cm^3 of this solution reacted exactly with 10.4 cm^3 of 0.0100 mol dm^{-3} potassium manganate(VII). What is the percentage by mass of iron in the tablets?

As with all titrations, we must consider **five** pieces of information:

- the balanced equation
  $$5Fe^{2+}(aq) + MnO_4^-(aq) + 8H^+(aq) \rightarrow 5Fe^{3+}(aq) + Mn^{2+}(aq) + 4H_2O(l)$$
- the concentration c_1 and reacting volume V_1 of $KMnO_4(aq)$
- the concentration c_2 and reacting volume V_2 of $Fe^{2+}(aq)$.

The concentration and volume of $KMnO_4$ are known so the amount of MnO_4^- ions can be found.

From the titration results, the amount of $KMnO_4$ can be calculated:

$$\text{amount of } KMnO_4 = c \times \frac{V}{1000} = 0.0100 \times \frac{10.4}{1000} = 1.04 \times 10^{-4} \text{ mol}$$

Using the equation determine the number of moles of the second reagent.

From the equation, the amount of Fe^{2+} can be determined:

$$5Fe^{2+}(aq) + MnO_4^-(aq) + 8H^+(aq) \rightarrow 5Fe^{3+}(aq) + Mn^{2+}(aq) + 4H_2O(l)$$

5 mol **1 mol** *(balancing numbers)*

$\therefore$ 5 x 1.04×10^{-4} mol Fe^{2+} reacts with 1.04×10^{-4} mol MnO_4^-

$\therefore$ amount of Fe^{2+} that reacted = 5.20×10^{-4} mol

If there is sampling of the original solution, you will need to scale.

Find the amount of Fe^{2+} in the solution prepared from the tablets:

10.0 cm^3 of $Fe^{2+}(aq)$ contains 5.20×10^{-4} mol Fe^{2+} (aq)

The 100 cm^3 solution of iron tablets contains $10 \times (5.20 \times 10^{-4})$
$= 5.20 \times 10^{-3}$ mol Fe^{2+}

Find the percentage of Fe^{2+} in the tablets (A_r: Fe, 55.8)

5.20×10^{-3} mol Fe^{2+} has a mass of $5.20 \times 10^{-3} \times 55.8 = 0.290$ g

$$\therefore \% \text{ of } Fe^{2+} \text{ in tablets} = \frac{\text{mass of } Fe^{2+}}{\text{mass of tablets}} \times 100 = \frac{0.290}{0.900} \times 100 = 32.2\%$$

For A2, you are expected to tackle unstructured titration calculations. The principles are the same, irrespective of the type of titration.

Further redox titrations

Providing there is a visible colour change, many other redox reactions can be used for redox titrations.

For example, acidified dichromate(VI) ions is an oxidising agent which can also oxidise Fe^{2+} ions to Fe^{3+}.

reduction: $Cr_2O_7^{2-}(aq) + 14H^+(aq) + 6e^- \longrightarrow 2Cr^{3+}(aq) + 7H_2O(l)$

oxidation: $Fe^{2+}(aq) \longrightarrow Fe^{3+}(aq) + e^-$

overall: $6Fe^{2+}(aq) + Cr_2O_7^{2-}(aq) + 14H^+(aq) \rightarrow 6Fe^{3+}(aq) + 2Cr^{3+}(aq) + 7H_2O(l)$

Iodine-thiosulfate titrations

OCR MS

Iodine-thiosulfate titrations can be used to determine iodine concentrations, either directly or by liberating iodine using a more powerful oxidising agent.

Iodine oxidises thiosulfate ions:

$$I_2(aq) + 2S_2O_3^{2-}(aq) \longrightarrow 2I^-(aq) + S_4O_6^{2-}(aq)$$

- Thiosulfate is added from the burette to the iodine solution until the deep-brown colour of iodine becomes a light-straw colour.
- Starch is now added, forming a deep-blue colour.
- Thiosulfate is added dropwise until the deep-blue colour becomes colourless. This is the end-point and shows that all the iodine has just been reacted.

Starch is a sensitive test for the presence of iodine. If added too early in this titration, the deep-blue colour perseveres and the end-point cannot be detected with accuracy.

The estimation of copper in a brass

1.65 g of a sample of brass was reacted with concentrated nitric acid. The solution was neutralised and diluted to 250 cm^3 with water.

25.0 cm^3 of this solution was pipetted into a flask. An excess of aqueous potassium iodide was then added to liberate iodine:

This is an unstructured titration calculation. See also pages 36–37 for general hints on solving this type of problem.

equation 1 $2Cu^{2+}(aq) + 4I^-(aq) \longrightarrow 2CuI(s) + I_2(s)$

Titration of this solution required 21.20 cm^3 of 0.100 mol dm^{-3} sodium thiosulfate.

equation 2 $2S_2O_3^{2-}(aq) + I_2(aq) \longrightarrow 2I^-(aq) + S_4O_6^{2-}(aq)$

Calculate the percentage, by mass, of copper in the brass sample.

From the titration results, the amount (in moles) of $Na_2S_2O_3$ can be calculated:

The concentration and volume of $Na_2S_2O_3$ are known.

$$\text{amount of } Na_2S_2O_3 = c \times \frac{V}{1000} = 0.100 \times \frac{21.20}{1000} = 0.00212 \text{ mol}$$

From the equations, the amount (in moles) of Cu^{2+} can be determined:

Using the equations, determine the number of moles of Cu^{2+}.

From *equation 2*, 2 mol $S_2O_3^{2-}$ reacts with 1 mol I_2

From *equation 1*, 1 mol I_2 produced from 2 mol Cu^{2+}

$\therefore$ 2 mol $S_2O_3^{2-}$ ≡ 1 mol I_2 ≡ 2 mol Cu^{2+}

$\therefore$ 0.00212 mol $S_2O_3^{2-}$ ≡ 0.00212/2 mol I_2 ≡ 0.00212 mol Cu^{2+}

amount of Cu^{2+} = 0.00212 mol

The amount of Cu^{2+} of the 250 cm^3 solution prepared from the brass can be deduced:

Don't forget to take into account any need to scale up.

25.0 cm^3 of $Cu^{2+}(aq)$ in the titration contains 0.00212 mol Cu^{2+}.

250 cm^3 of the $Cu^{2+}(aq)$ solution from the brass contains 10 x 0.00212 = 0.0212 mol Cu^{2+}.

Finally, the percentage, by mass, of copper in brass can be calculated:

The final step is now easy provided that you have remembered:

$$n = \frac{\text{mass}}{\text{molar mass}}$$

0.0212 mol Cu produced 0.0212 mol Cu^{2+}

mass of copper in brass = 63.5 x 0.0212 = 1.35 g

$$\therefore \text{percentage of copper in brass} = \frac{1.35}{1.65} \times 100 = 81.8\ \%.$$

Progress check

1 Write a full equation from the following pairs of half-equations. For each pair, identify the changes in oxidation number, what has been oxidised and what has been reduced.

(a) $Zn \longrightarrow Zn^{2+} + 2e^-$

$VO_2^+ + 4H^+ + 3e^- \longrightarrow V^{2+} + 2H_2O$

(b) $Cr^{3+} + 8OH^- \longrightarrow CrO_4^{2-} + 4H_2O + 3e^-$

$H_2O_2 + 2e^- \longrightarrow 2OH^-$

2 In a redox titration, 25.0 cm^3 of an acidified solution containing $Fe^{2+}(aq)$ ions reacted exactly with 21.8 cm^3 of 0.0200 mol dm^{-3} potassium manganate(VII). Calculate the concentration of $Fe^{2+}(aq)$ ions.

1 (a) $2VO_2^+ + 3Zn + 8H^+ \longrightarrow 2V^{2+} + 3Zn^{2+} + 4H_2O$

V: +5 → +2; Zn: 0 → +2; VO_2^+ reduced; Zn oxidised.

(b) $2Cr^{3+} + 3H_2O_2 + 10OH^- \longrightarrow 2CrO_4^{2-} + 8H_2O$

Cr: +3 → +6; O: −1 → −2; Cr^{3+} oxidised; H_2O_2 reduced.

2 0.0872 mol dm^{-3}

3.5 Catalysis

After studying this section you should be able to:

- *understand how a transition element acts as a catalyst*
- *explain how a catalyst acts in heterogeneous catalysis*
- *explain how a catalyst acts in homogeneous catalysis*

LEARNING SUMMARY

Key points from AS

- **How do catalysts work?** *Revise AS page 86–87*
- **Heterogeneous and homogeneous catalysis** *Revise AS pages 86–88*

The two different classes of catalyst, homogeneous and heterogeneous, were discussed in detail during AS Chemistry. Both types are used industrially but heterogeneous catalysis is much more common. The AS coverage is summarised below and expanded to include extra examples of transition element catalysis.

Transition elements as catalysts

OCR M5
SALTERS M4

An important use of transition metals and their compounds is as catalysts for many industrial processes. Nickel and platinum are extensively used in the petroleum and polymer industries.

> Transition metal ions are able to act as catalysts by changing their oxidation states. This is made possible by using the partially full d-orbitals for gaining or losing electrons.
>
> KEY POINT

Heterogeneous catalysis

SALTERS M4

Key points from AS

- **Heterogeneous and homogeneous catalysis** *Revise AS pages 86–88*

A **heterogeneous** catalyst has a **different** phase from the reactants.

- Many examples of heterogeneous catalysts involve reactions between **gases**, catalysed by a **solid** catalyst which is often a transition metal or one of its compounds.
- Transition metals use d and s-electrons to form weak bonds with reactant molecules at the catalyst surface.

The process involves:

- **diffusion** of gas molecules onto the surface of the iron catalyst
- **adsorption** of the gases to the surface of the catalyst
- **weakening of bonds**, in reactant molecules
- **bond breaking**, in reactant molecules
- **bond making**, in product molecules
- **diffusion** of the product molecules from the surface of the catalyst, allowing more gas molecules to diffuse onto its surface.

Transition metal ions as heterogeneous catalysts

The catalytic converter

- Rh/Pt/Pd catalysts are used in catalytic converters for the removal of polluting gases, such as CO and NO, produced in a car engine.
- The **solid** Rh/Pt/Pd catalyst is held on a ceramic honeycomb support, giving a large surface area for the catalyst which can be spread extremely thinly on the support. The high surface area of the catalyst increases the chances of a reaction and less of the expensive catalysts are needed, keeping down costs.

Iron

Iron-containing catalysts are used in many industrial processes, the most important being to form ammonia from nitrogen and hydrogen in the Haber process.

$$N_2(g) \ + \ 3H_2(g) \rightleftharpoons 2NH_3(g)$$

Chromium(III) oxide

Chromium(III) oxide, Cr_2O_3, is used to catalyse the manufacture of methanol from carbon monoxide and hydrogen.

$$CO(g) \ + \ 2H_2(g) \rightleftharpoons CH_3OH(g)$$

Vanadium

A vanadium(V) oxide, V_2O_5, catalyst is used in the contact process for sulfur trioxide production from which sulfuric acid is manufactured.

This catalysis proceeds via an **intermediate state**.

$$SO_2(g) + \tfrac{1}{2}O_2(g) \rightleftharpoons SO_3(g) \qquad \textit{The contact process}$$

Homogeneous catalysis

SALTERS M4

Key points from AS

- **How catalysts work**
  Revise AS page 86–87
- **Homogeneous catalysis**
  Revise AS page 87–89

A **homogeneous** catalyst has the **same** phase as the reactants.

- During homogeneous catalysis, the catalyst initially reacts with the reactants forming an **intermediate state** with **lower activation energy**.
- The catalyst is then regenerated as the reaction completes, allowing the catalyst to react with further reactants.
- Overall the catalyst is **not** used up, but it is actively involved in the process by a chain reaction.

Transition metal ions as homogeneous catalysts

The reaction between I^- and $S_2O_8^{2-}$, catalysed by Fe^{2+}(aq) ions

The reaction between I^- and $S_2O_8^{2-}$ is very slow:

- it can be catalysed by Fe^{2+} ions
- the Fe^{2+} ions are recycled in a chain reaction.

Transition metal ions often change their oxidation states during their catalytic action.
Here: $Fe^{2+} \longrightarrow Fe^{3+}$ (and back to Fe^{2+})

chain reaction:

$$S_2O_8^{2-}(aq) + 2Fe^{2+}(aq) \longrightarrow 2SO_4^{2-}(aq) + 2Fe^{3+}(aq)$$

$$2Fe^{3+}(aq) + 2I^-(aq) \longrightarrow I_2(aq) + 2Fe^{2+}(aq)$$

overall:

$$S_2O_8^{2-}(aq) + 2I^-(aq) \longrightarrow I_2(aq) + 2SO_4^{2-}(aq)$$

The overall reaction between I^- and $S_2O_8^{2-}$ involves two negative ions. These negative ions will repel one another, making it difficult for a reaction to take place. Notice that, in each stage of the mechanism, the catalyst provides positive ions which attract, and react with, each negative ion in turn.

Progress check

1 (a) State what is meant by homogeneous and heterogeneous catalysis.
(b) State an example of a homogeneous and heterogeneous catalyst.

2 How does a transition metal act as a catalyst?

1 (a) Homogeneous: catalyst and reactants have the same phase.
Heterogeneous: catalyst and reactants have different phases.
(b) Homogeneous: Fe^{2+} catalysing the reaction between I^- and $S_2O_8^{2-}$.
Heterogeneous: iron catalysing the reaction between N_2 and H_2 for ammonia production (Haber process).

2 Transition metal changes oxidation states by gaining or losing electrons from partially filled d-orbitals.

3.6 Reactions of metal aqua-ions

After studying this section you should be able to:

- *describe the precipitation reactions of metal aqua-ions with bases*
- *describe the acidity of transition metal aqua-ions*

LEARNING SUMMARY

Simple precipitation reactions of metal aqua-ions

OCR M5
SALTERS M4

A precipitation reaction takes place between **aqueous alkali** and an aqueous solution of a **metal(II)** or **metal(III) cation**.

This results in the formation of a **precipitate** of the **metal hydroxide**, often with a characteristic colour.

Notice that NaOH(aq) and NH_3(aq) **both** act as bases.

Suitable aqueous alkalis include:

- aqueous sodium hydroxide, NaOH(aq)
- aqueous ammonia, NH_3(aq).

Aqueous ammonia provides OH^-(aq) ions.

$NH_3(aq) + H_2O(l) \rightleftharpoons NH_4^+(aq) + OH^-(aq)$

The hydroxides of all transition metals are insoluble in water.

These precipitation reactions can be represented simply as follows.

$$Cu^{2+}(aq) + 2OH^-(aq) \longrightarrow Cu(OH)_2(s)$$
$$Cr^{3+}(aq) + 3OH^-(aq) \longrightarrow Cr(OH)_3(s)$$

The characteristic colour of the precipitate can help to identify the metal ion. The colours of some hydroxide precipitates are shown below.

hydroxide	***colour***
$Fe(OH)_2(s)$	green
$Fe(OH)_3(s)$	brown
$Co(OH)_2(s)$	blue–green
$Cu(OH)_2(s)$	blue
$Cr(OH)_3(s)$	green

Reaction of complex metal aqua-ions with aqueous alkali

See also: 'Reactions of metal aqua-ions with aqueous bases' page 85.

Equations can be written using complex aqua-ions for the reactions above, each producing a **precipitate** of the **hydrated hydroxide**.

The precipitate has no charge.

$$[Cu(H_2O)_6]^{2+}(aq) + 2OH^-(aq) \longrightarrow Cu(OH)_2(H_2O)_4(s) + 2H_2O(l)$$

blue precipitate

$$[Cr(H_2O)_6]^{3+}(aq) + 3OH^-(aq) \longrightarrow Cr(OH)_3(H_2O)_3(s) + 3H_2O(l)$$

green precipitate

In exams you can show hydroxide precipitates based on six ligands, e.g. $Cu(OH)_2(H_2O)_4(s)$ – or you can use simplified formulae such as $Cu(OH)_2(s)$.

Species in solution are charged.

Excess aqueous ammonia

Excess ammonia usually results in **ligand exchange** forming a soluble ammine complex:

Salters prefer to use $[Cu(NH_3)_4]^{2+}$ for the complex formed between Cu^{2+} and NH_3.

$$Cr(H_2O)_3(OH)_3(s) + 6NH_3(aq) \longrightarrow [Cr(NH_3)_6]^{3+}(aq) + 3H_2O(l) + 3OH^-(aq)$$

purple solution

If aqueous ammonia is slowly added to an aqueous solution of a **metal(II)** or **metal(III) cation**:

- a precipitate of the metal hydroxide first forms: **acid–base reaction**
- the precipitate then dissolves in excess ammonia: **ligand substitution.**

Reactions of metal aqua-ions with aqueous bases

OCR M5
SALTERS M4

Reactions with $OH^-(aq)$ and $NH_3(aq)$

You are expected to know reactions of metal aqua ions with aqueous $OH^-(aq)$ ions (probably as NaOH(aq)) and with aqueous ammonia, $NH_3(aq)$.

The tables below summarise what you are expected to know for the two Boards so make sure that you do not learn more than you have to!

In exams, a great emphasis is placed on these reactions. You will need to learn the colours of all solutions and precipitates in your course.

	OCR
Fe^{2+}	✓
Co^{2+}	✓
Cu^{2+}	✓
Fe^{3+}	✓
OH^-	✓
NH_3	✓
NH_3 excess	✓

	SALTERS
Fe^{2+}	✓
Co^{2+}	
Cu^{2+}	✓
Fe^{3+}	✓
OH^-	✓
NH_3	✓
NH_3 excess	✓

	$OH^-(aq)$ or $NH_3(aq)$ as a base	***$NH_3(aq)$ as a ligand***
aqua ion	***NaOH(aq) or dilute $NH_3(aq)$ (small amount)***	***excess/conc. $NH_3(aq)$***
$[Fe(H_2O)_6]^{2+}$ green	$Fe(OH)_2(s)$ / $Fe(OH)_2(H_2O)_4(s)$ green	
$[Co(H_2O)_6]^{2+}$ pink	$Co(OH)_2(s)$ / $Co(OH)_2(H_2O)_4(s)$ blue–green	
$[Cu(H_2O)_6]^{2+}$ blue	$Cu(OH)_2(s)$ / $Cu(OH)_2(H_2O)_4(s)$ pale blue	$[Cu(NH_3)_4(H_2O)_2]^{2+}$ deep blue
$[Fe(H_2O)_6]^{3+}$ red-brown	$Fe(OH)_3(s)$ / $Fe(OH)_3(H_2O)_3(s)$ brown	

- $Fe(OH)_2(s)$ is oxidised in air forming a brown precipitate of $Fe(OH)_3(s)$.

Progress check

1 Write equations for the following reactions of metal aqua-ions with aqueous alkali. What is the colour of each precipitate?

(a) Co^{2+}(aq) (b) Fe^{2+}(aq) (c) $[Ni(H_2O)_6]^{2+}$ (d) $[Fe(H_2O)_6]^{3+}$.

2 Write equations to show what happens when excess NH_3(aq) is added to the following metal aqua ions.

(a) $[Co(H_2O)_6]^{2+}$ (b) $[Cu(H_2O)_6]^{2+}$ (c) $[Cr(H_2O)_6]^{3+}$.

1 (a) $Co^{2+}(aq) + 2OH^-(aq) \longrightarrow Co(OH)_2(s)$ (blue–green)
(b) $Fe^{2+}(aq) + 2OH^-(aq) \longrightarrow Fe(OH)_2(s)$ (green)
(c) $[Ni(H_2O)_6]^{2+}(aq) + 2OH^-(aq) \longrightarrow Ni(OH)_2(H_2O)_4(s) + 2H_2O(l)$ (pale green)
(d) $[Fe(H_2O)_6]^{3+}(aq) + 3OH^-(aq) \longrightarrow Fe(OH)_3(H_2O)_3(s) + 3H_2O(l)$ (brown)

2 (a) $[Co(H_2O)_6]^{2+}(aq) + 6NH_3(aq) \longrightarrow [Co(NH_3)_6]^{2+}(aq) + 6H_2O(l)$ (pale straw)
(b) $[Cu(H_2O)_6]^{2+}(aq) + 4NH_3(aq) \longrightarrow [Cu(NH_3)_4(H_2O)_2]^{2+}(aq) + 4H_2O(l)$ (deep-blue)
(c) $[Cr(H_2O)_6]^{3+}(aq) + 6NH_3(aq) \longrightarrow [Cr(NH_3)_6]^{3+}(aq) + 6H_2O(l)$ (purple)

Sample question and model answer

Remember:
4s fill before 3d
4s empty before 3d

This question looks at different transition elements.

(a) Write the electron configuration of a vanadium atom and a cobalt(III) ion.

V: $1s^2 2s^2 2p^6 3s^2 3p^6 3d^3 4s^2$ ✓

Co^{3+}: $1s^2 2s^2 2p^6 3s^2 3p^6 3d^6$ ✓ [2]

This is standard bookwork but it also reinforces important principles:
H_2O forms octahedral complexes with a coordination number of 6.
Cl^- forms tetrahedral complexes with a coordination number of 4.

(b) When concentrated hydrochloric acid is added to an aqueous solution containing copper(II) ions, the colour changes. Write the formulae of the two copper complexes involved, their colours, their shapes and the co-ordination numbers of the copper. Write an equation for the reaction and explain why the addition of dilute hydrochloric acid does not produce the same colour change.

Aqueous copper(II) contains the complex $[Cu(H_2O)_6]^{2+}$ ✓ which is pale blue. ✓ It has an octahedral shape ✓ and a coordination number of 6. ✓

After addition of concentrated hydrochloric acid, the complex ion $[CuCl_4]^{2-}$ ✓ is formed, which is yellow. ✓ It has a tetrahedral shape ✓ and a coordination number of 4. ✓

Equation: $[Cu(H_2O)_6]^{2+} + 4Cl^- \longrightarrow CuCl_4^{2-} + 6H_2O$ ✓✓

The concentration of chloride ligands in **dilute** hydrochloric acid is much smaller than the concentration of water ligands. ✓ [11]

Learn how to combine redox half-equations. Although this titration is different from the permanganate titrations that you have carried out, the same principles are used.
Even if your equation is wrong, you can still score in the next part provided that your method is built upon sound chemical principles.
Show your working!

(c) A 0.626 g sample of hydrated iron(II) sulfate was dissolved in water and the solution made up to 250 cm^3. A 25.0 cm^3 sample of this solution was acidified with dilute sulfuric acid and titrated with potassium dichromate(VI). 25.85 cm^3 of 0.00145 mol dm^{-3} potassium dichromate(VI) were required for exact reaction.

The half-equation for the reduction of acidified dichromate ions is shown below.

$$Cr_2O_7^{2-}(aq) + 14H^+(aq) + 6e^- \longrightarrow 2Cr^{3+}(aq) + 7H_2O(l)$$

(i) Write the half-equation for the oxidation of Fe^{2+} and construct the equation for the overall reaction with acidified dichromate (VI) ions.

$Fe^{2+}(aq) \longrightarrow Fe^{3+}(aq) + e^-$ ✓

$6Fe^{2+}(aq) + Cr_2O_7^{2-}(aq) + 14H^+(aq) \longrightarrow 6Fe^{3+}(aq) + 2Cr^{3+}(aq) + 7H_2O(l)$ ✓

(ii) Using the information above, calculate the molar mass of the hydrated iron(II) sulfate and determine the number of moles of water of crystallisation present in one mole of hydrated iron(II) sulfate.

amount of $Cr_2O_7^{2-} = c \times \frac{V}{1000} = 0.00145 \times \frac{25.85}{1000} = 3.75 \times 10^{-5}$ mol ✓

From the equation, the amount of $Fe^{2+} = 6 \times$ amount of $Cr_2O_7^{2-}$. ✓

∴ amount of Fe^{2+} that reacted $= 6 \times 3.75 \times 10^{-5}$ mol $= 2.25 \times 10^{-4}$ mol ✓

Don't forget the scaling stage. The original solution was 250 cm^3 but you have only taken 25 cm^3 for the titration.
Show your working!

In 250 cm^3 solution, amount of $Fe^{2+} = 10 \times 2.25 \times 10^{-4} = 2.25 \times 10^{-3}$ mol ✓

∴ $\frac{0.626}{2.25 \times 10^{-3}} = 278$ g of hydrated iron(II) sulfate contains 1 mol Fe^{2+} ✓

M of hydrated iron(II) sulfate = 278 g mol^{-1} ✓

M of $FeSO_4 = 55.8 + 32.1 + 16.0 \times 4 = 151.9$ g mol^{-1} ✓

Number of waters of crystallisation $= \frac{278.0 - 151.9}{18.0} = \frac{126.1}{18.0} = 7$ ✓

(iii) Explain how the volume of potassium dichromate(VII) solution needed in the titration would be different if the solution of iron(II) sulfate had been left for some time before being titrated.

Less dichromate(VI) solution would have been needed ✓ because some of the Fe^{2+} would have been oxidised by air to Fe^{3+} ✓ [12]

[Total: 25]

Practice examination questions

1 (a) Write the electron configuration of a nickel atom and a Ni^{2+} ion.[2]

(b) Nickel is both a d-block element and a transition element. State what is meant by each of the terms. [2]

(c) In aqueous solution, nickel sulfate, forms a six-coordinate complex ion which gives a green colour to the solution.

(i) What feature of the water molecule allows it to form a complex ion with Ni^{2+}?

(ii) Write the formula of the nickel complex ion formed in aqueous solution.

(iii) What types of bond are present in this nickel complex ion?

(iv) Suggest the shape of, and bond angles in, this complex ion. [6]

(d) Consider the following reactions.

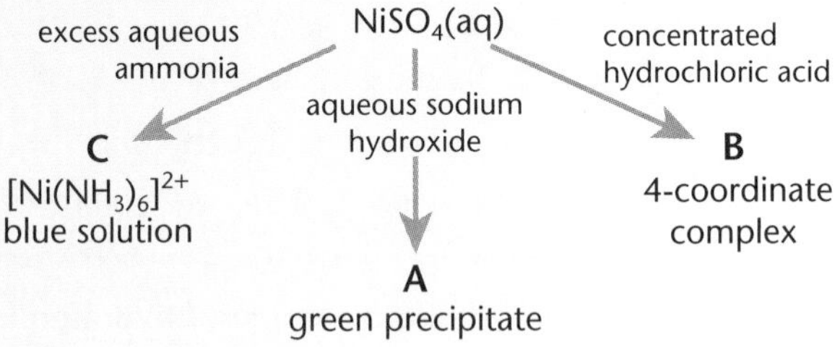

(i) Write the formula of the green precipitate, **A**, and write an ionic equation for its formation above.

(ii) Predict the formula of **B**.

(iii) Write an equation for the formation of $[Ni(NH_3)_6]^{2+}$ from the aqueous nickel complex ion and suggest the type of reaction taking place. [5]

[Total: 15]

2 (a) Deduce the oxidation state of the transition metal in each of the following species.

(i) Cr in CrO_4^{2-}

(ii) Cu in $[CuCl_2]^-$

(iii) V in $[VO(H_2O)_5]^{2+}$ [3]

(b) Chromium(III) can be oxidised in alkaline conditions by hydrogen peroxide, H_2O_2. Use the half-equations below to construct the overall equation for this reaction.

$$Cr^{3+} + 8OH^- \longrightarrow CrO_4^{2-} + 4H_2O + 3e^-$$

$$H_2O_2 + 2e^- \longrightarrow 2OH^-$$ [1]

(c) Give two examples of the use of transition elements as catalysts in industrial processes. [2]

[Total: 6]

Practice examination questions (continued)

3 The concentration of an aqueous solution of hydrogen peroxide can be determined by titration. Aqueous potassium manganate(VII), $KMnO_4$, is titrated against a solution of hydrogen peroxide in the presence of acid.

The half-equations are shown below.

$$H_2O_2(aq) \longrightarrow 2H^+(aq) + O_2(g) + 2e^-$$

$$MnO_4^-(aq) + 8H^+(aq) + 5e^- \longrightarrow Mn^{2+}(aq) + 4H_2O(l)$$

(a) (i) Deduce the oxidation state of the manganese in the MnO_4^- ion.

(ii) Construct the equation for the reaction between H_2O_2, MnO_4^- ions and H^+ ions. [2]

(b) State the role of hydrogen peroxide in this reaction. [1]

(c) After acidification, 25.0 cm^3 of a dilute aqueous solution of hydrogen peroxide reacted exactly with 23.80 cm^3 of 0.0150 mol dm^{-3} $KMnO_4$. Calculate the concentration, in mol dm^{-3}, of hydrogen peroxide in the solution. [3]

[Total: 6]

OCR

4 Cobalt and nickel both form complex ions with the ligand ethane-1,2-diamine, $NH_2CH_2CH_2NH_2$.

(a) (i) Explain how ligands bond to a metal ion in a complex ion.

(ii) What name is given to this type of ligand? [3]

(b) Cobalt forms the complex compound, $[Co(NH_2CH_2CH_2NH_2)_3]Cl_3$.

(i) What is the oxidation state of cobalt in this compound?

(ii) What is the electron configuration of cobalt in this compound? [2]

(c) In solution, nickel forms the complex ion $[Ni(NH_2CH_2CH_2NH_2)_3]^{2+}$.

(i) Deduce the co-ordination number of nickel in this complex ion.

(ii) State the shape around nickel in this complex ion.

(iii) Outline how you could prepare a solution containing the complex ion $[Ni(NH_2CH_2CH_2NH_2)_3]^{2+}$ from an aqueous solution of nickel(II) chloride. Include an equation in your answer. [4]

[Total: 9]

Chapter 4

Chemistry of organic functional groups

The following topics are covered in this chapter:

- *Isomerism and functional groups*
- *Aldehydes and ketones*
- *Carboxylic acids*
- *Esters*
- *Acylation*
- *Arenes*
- *Reactions of arenes*
- *Phenols and alcohols*
- *Amines*

4.1 Isomerism and functional groups

LEARNING SUMMARY

After studying this section you should be able to:

- *understand what is meant by structural isomerism and stereoisomerism*
- *explain the term chiral centre and identify any chiral centres in a molecule of given structural formula*
- *understand that many natural compounds are present as one optical isomer only*
- *understand that many pharmaceuticals are chiral drugs*
- *recognise common functional groups*

Key points from AS

- **Basic concepts**
  Revise AS pages 97–102

During AS Chemistry, you learnt about the basic concepts used in organic chemistry. You should revise these thoroughly before you start the A2 part of this course.

Isomerism

OCR M4
SALTERS M4, M5

Structural and *E/Z* isomerism were introduced during AS Chemistry. These are reviewed below and *E/Z* isomerism is discussed in the wider context of **stereoisomerism**.

Key points from AS

- **Structural isomerism**
  Revise AS pages 99–100
- ***E/Z* isomerism**
  Revise AS pages 107–108

Structural isomerism

Structural isomers are molecules with the same molecular formula but with different structural arrangements of atoms (structural formulae).

Two structural isomers of $C_2H_4O_2$ are shown below:

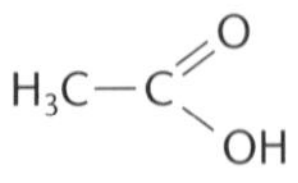

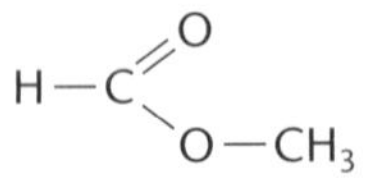

Stereoisomerism

Stereoisomers are molecules with the same structural formula but their atoms have different positions in space.

There are two types of stereoisomerism, each arising from a different structural feature:

- *E/Z* isomerism about a **C=C** double bond
- **optical** isomerism about a **chiral** carbon centre.

In simple cases in which 2 of the groups are the same, *E/Z* isomerism is often referred to as *cis-trans* isomerism.

Each *E/Z* isomer has the same structural formula.

E/Z isomerism

E/Z isomerism occurs in molecules with:

- a C=C double bond and
- two **different** groups attached to **each** carbon in the C=C bond.

The double bond prevents rotation.

E.g. *E/Z* isomers of 1,2-dichloroethene, ClCH=CHCl.

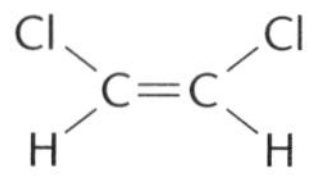

***cis*-isomer**
(groups on one side)
Z isomer

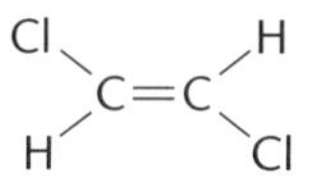

***trans*-isomer**
(groups on opposite sides)
E isomer

Optical isomers

> Optical isomerism occurs in the molecules of a compound with a **chiral** (or *asymmetric*) carbon atom. A chiral carbon atom has **four** different groups attached to it.
>
> **KEY POINT**

Optical isomers are usually drawn as 3D diagrams. It is then easy to picture the mirror images.

Optical isomers are also called **enantiomers**.

E.g. the optical isomers of an amino acid, $RCHNH_2COOH$ (see pages 127–129).

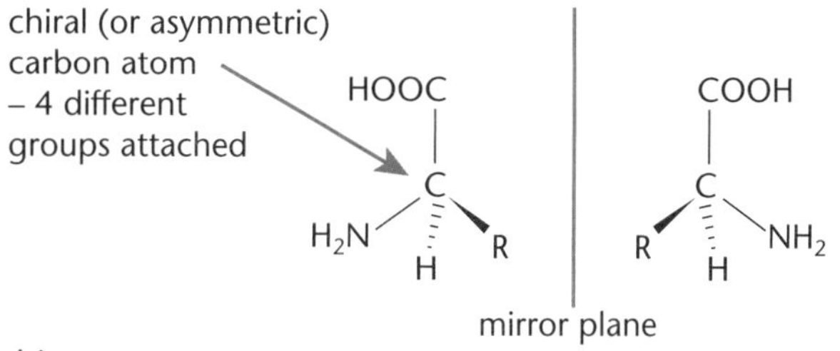

Optical isomers:

- are non-superimposable mirror images of one another
- rotate plane-polarised light in opposite directions
- are chemically identical.

Society discovered the consequences of harmful side-effects from the 'wrong' optical isomer with the use of thalidomide. One optical isomer combated the effects of morning sickness in pregnant women. The other optical isomer was the cause of deformed limbs of unborn babies.

Partly through this lesson, drugs are now often used as the optically pure form, comprising just the required optical isomer.

Chirality and drug synthesis

A synthetic amino acid, made in the laboratory, is optically inactive and does not rotate plane-polarised light:

- it contains equal amounts of each optical isomer – a **racemic** mixture or **racemate**.

A natural amino acid, made by living systems, is optically active:

- it contains only one of the optical isomers.

The difference between the optical isomers present in natural and synthetic organic molecules has important consequences for drug design. The synthesis of pharmaceuticals often requires the production of chiral drugs containing a single optical isomer. Although one of the optical isomers may have beneficial effects, the other may be harmful and may lead to undesirable side effects. See also page 155.

Progress check

1 Show the alkenes that are structural isomers of C_4H_8.

2 Show the *E/Z* isomers of C_4H_8.

3 Show the optical isomers of C_4H_9OH.

1 $CH_3CH_2CH{=}CH_2$; $CH_3CH{=}CHCH_3$; $(CH_3)_2C{=}CH_2$

2

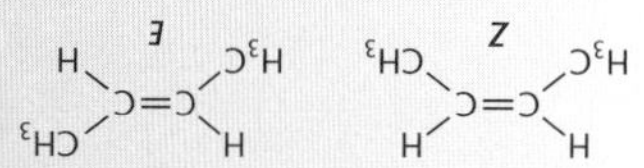

3

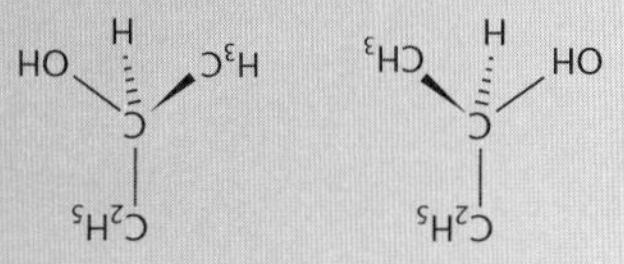

Common functional groups

OCR	M4
SALTERS	M4, M5

It is essential that you can instantly identify a functional group within a molecule so that you can apply the relevant chemistry.

You must learn all of these!

The functional groups in the table below contain most of the functional groups you will meet in the A Level Chemistry course, including those met in AS Chemistry. It cannot be stressed enough just how important it is that you can instantly recognise a functional group. Without this, your progress in Organic Chemistry will be very limited.

name	functional group	examples	structural formula	prefix or suffix (for naming)
alkane	C–H	$CH_3CH_2CH_3$ *propane*	$CH_3CH_2CH_3$	-ane
alkene	$>C{=}C<$	CH_3CHCH_2 *propene*	$(H_3C)(H)C{=}C(H)(H)$	-ene
halogenoalkane	— Br	CH_3CH_2Br *bromoethane*	CH_3CH_2—Br	bromo-
alcohol	— OH	CH_3CH_2OH *ethanol*	CH_3CH_2—OH	-ol
aldehyde	$-C({=}O)H$	CH_3CHO *ethanal*	$H_3C-C({=}O)H$	-al
ketone	$-C({=}O)-$	CH_3COCH_3 *propanone*	$H_3C-C({=}O)CH_3$	-one
carboxylic acid	$-C({=}O)OH$	CH_3COOH *ethanoic acid*	$H_3C-C({=}O)OH$	-oic acid
ester	$-C({=}O)O-$	CH_3COOCH_3 *methyl ethanoate*	$H_3C-C({=}O)O-CH_3$	-oate
acyl chloride	$-C({=}O)Cl$	CH_3COCl *ethanoyl chloride*	$H_3C-C({=}O)Cl$	–oyl chloride
amine	— NH_2	$CH_3CH_2NH_2$ *ethylamine*	CH_3CH_2—NH_2	-amine
amide	$-C({=}O)NH_2$	CH_3CONH_2 *ethanamide*	$H_3C-C({=}O)NH_2$	-amide
nitrile	$-C{\equiv}N$	CH_3CN *ethanenitrile*	H_3C—CN	-nitrile

Key points from AS

- **Functional groups**
  Revise AS pages 98–99

The nitrile carbon atom is included in the name.

CH_3CN contains the longest carbon chain with **two** carbon atoms and its name is based upon ethane – hence ethanenitrile.

4.2 Aldehydes and ketones

LEARNING SUMMARY

After studying this section you should be able to:

- *understand the polarity and physical properties of carbonyl compounds*
- *describe the oxidation of alcohols*
- *describe the reduction of carbonyl compounds to form alcohols*
- *describe nucleophilic addition to aldehydes and ketones*
- *describe a test to detect the presence of the carbonyl group*
- *describe tests to detect the presence of an aldehyde group*

Carbonyl compounds General formula: $C_nH_{2n}O$

OCR	M4
SALTERS	M4

During AS Chemistry, you learnt about how alcohols can be oxidised to carbonyl compounds: aldehydes and ketones. For A2 Chemistry, you will learn about the reactions of aldehydes and ketones.

Key points from AS

- **Oxidation of alcohols**
  Revise AS pages 113–114

Types and naming of carbonyl compounds

The carbonyl group, C=O, is the functional group in aldehydes and ketones.

aldehyde, RCHO

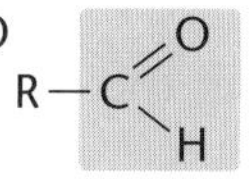

ketone, RCOR'

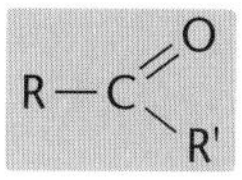

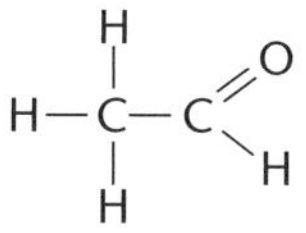

ethanal
CH_3CHO

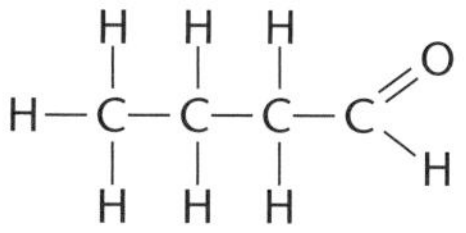

butanal
$CH_3CH_2CH_2CHO$

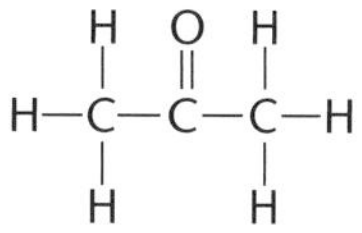

propanone
CH_3COCH_3

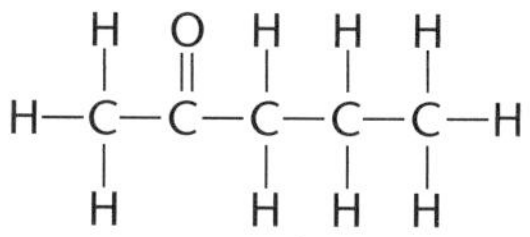

pentan-2-one
$CH_3COCH_2CH_2CH_3$

Polarity of carbonyl compounds

Carbon and oxygen have different electronegativities, resulting in a polar C=O bond: carbonyl compounds have polar molecules.

The properties of aldehydes and ketones are dominated by the polar carbonyl group, C=O.

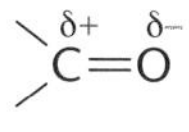

oxygen is more electronegative than carbon producing a dipole

Physical properties of carbonyl compounds

The polarity of the carbonyl group is less than that of the hydroxyl group in alcohols. Thus, aldehydes and ketones have weaker dipole–dipole interactions and lower boiling points than alcohols of comparable molecular mass.

The polarity in propanone is such that it mixes with polar solvents such as water and also dissolves many organic compounds. The low boiling point also makes it easy to remove by evaporation, a property exploited by its use in paints and varnishes.

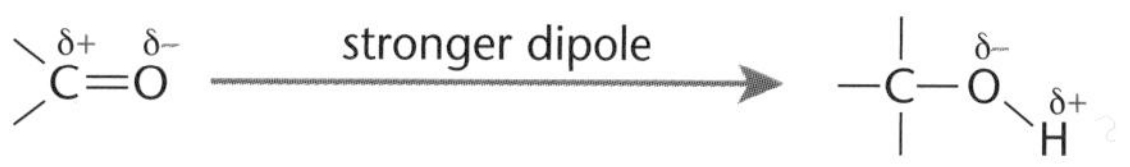

carbonyl group in aldehyde or ketone

hydroxyl group in alcohols

Oxidation of alcohols

OCR M4
SALTERS M4

Key points from AS

- **Oxidation of alcohols**
  Revise AS pages 113–114

You studied the oxidation of different types of alcohols during AS Chemistry. The link between alcohols, carbonyl compounds and carboxylic acids is shown below.

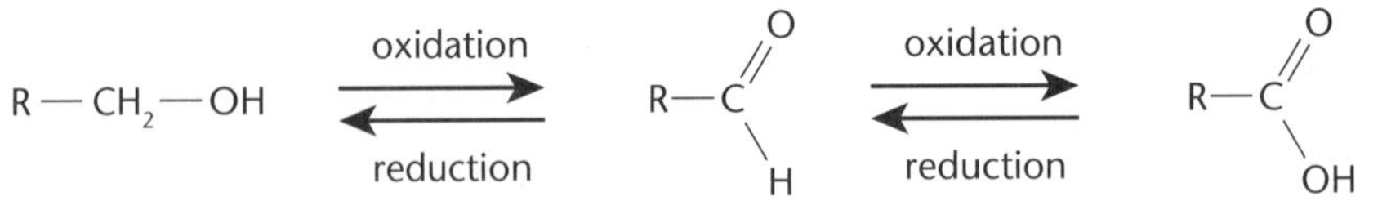

primary alcohol   aldehyde   carboxylic acid

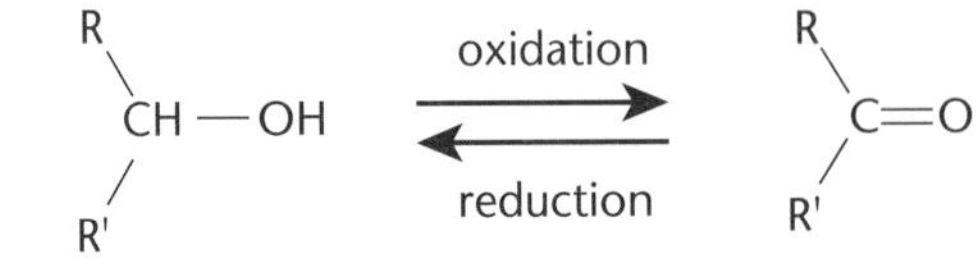

secondary alcohol   ketone

Oxidation of primary alcohols

The orange dichromate ions, $Cr_2O_7^{2-}$, are reduced to green Cr^{3+} ions.

A primary alcohol can be oxidised to an aldehyde and then to a carboxylic acid.

This is carried out using an oxidising agent such as a mixture of concentrated sulfuric acid, H_2SO_4 (source of H^+) and potassium dichromate, $K_2Cr_2O_7$ (source of $Cr_2O_7^{2-}$).

- By heating and distilling the product immediately, oxidation can be stopped at the aldehyde stage.

$$CH_3CH_2OH + [O] \longrightarrow CH_3CHO + H_2O$$

For balanced equations, the oxidising agent can be shown simply as [O].

- By refluxing with an excess of the oxidising agent, further oxidation takes place to form the carboxylic acid.

$$CH_3CH_2OH + 2[O] \longrightarrow CH_3COOH + H_2O$$

Oxidation of secondary alcohols

By heating with $H^+/Cr_2O_7^{2-}$, a secondary alcohol can be oxidised to a ketone. No further oxidation normally takes place.

The equation for the oxidation of propan-2-ol is shown below.

$$CH_3(CHOH)CH_3 + [O] \longrightarrow CH_3COCH_3 + H_2O$$

Reduction of carbonyl compounds

OCR M4

Aldehydes and ketones can be reduced to alcohols using a reducing agent containing the hydride ion, H^-.

Suitable reducing agents are:

- sodium tetrahydridoborate(III) (*sodium borohydride*), $NaBH_4$, in water (heat)
- lithium tetrahydridoaluminate(III) (*lithium aluminium hydride*), $LiAlH_4$, in ether.

Aldehydes are reduced to primary alcohols:

For balanced equations, the reducing agent can be shown simply as [H].

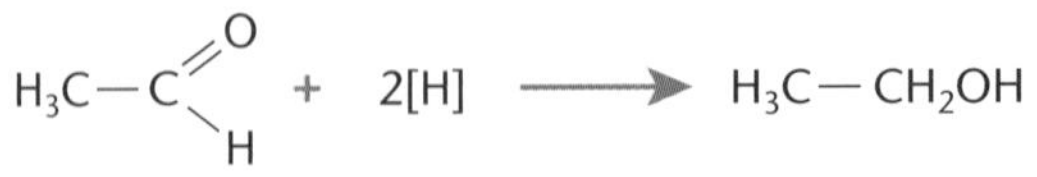

aldehyde   primary alcohol

$NaBH_4$ and $LiAlH_4$ both reduce the C=O double bond in aldehydes and ketones.

They do **not** reduce the C=C bond in alkenes.

Ketones are reduced to secondary alcohols:

ketone   secondary alcohol

Nucleophilic addition to carbonyl compounds

OCR M4
SALTERS M4

The electron-deficient carbon atom of the polar $C^{\delta+}{=}O^{\delta}$ bond attracts **nucleophiles**. This allows an **addition** reaction to take place across the C=O double bond of aldehydes and ketones. This is called **nucleophilic addition.**

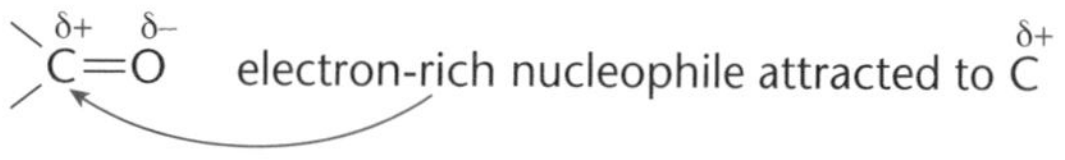

OCR M4

The reaction of carbonyl compounds with hydrogen cyanide is also an example of nucleophilic addition.

Reduction as nucleophilic addition

The reduction of aldehydes and ketones to alcohols using $NaBH_4$ or $LiAlH_4$ is an example of **nucleophilic addition**. The reducing agent can be considered to release hydride ions, H^-.

$$NaBH_4 \longrightarrow H^- + NaBH_3^+$$

The hydride ion acts as a nucleophile in the first stage of the reaction.

Mechanism

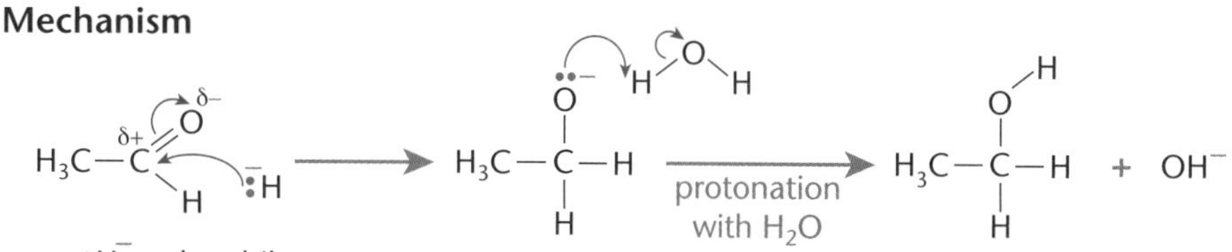

:H⁻ nucleophile donates electron pair

primary alcohol

SALTERS M4

HCN is added across the C=O double bond.

Nucleophilic addition of hydrogen cyanide, HCN

In the presence of cyanide ions, CN^-, hydrogen cyanide, HCN, is added across the C=O bond in aldehydes and ketones.

aldehyde	+	HCN ⟶	hydroxynitrile
CH_3CHO	+	HCN ⟶	$CH_3CH(OH)CN$

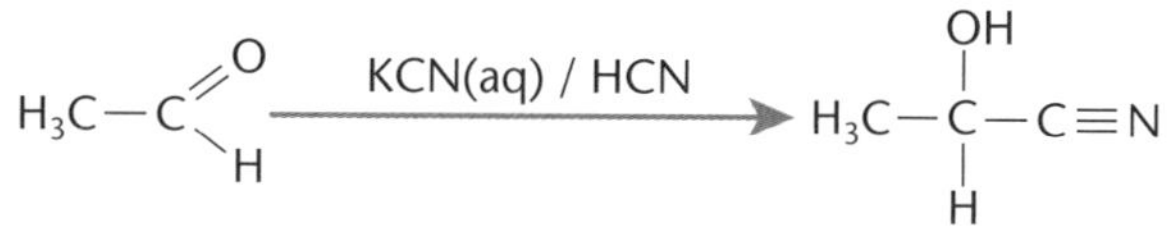

This hydroxynitrile has optical isomers.

Hydrogen cyanide is a very poisonous gas and it is usually generated in solution as H^+ and CN^- ions using:

- sodium cyanide, NaCN as a source of CN^-
- dilute sulfuric acid as a source of H^+.

The presence of cyanide ions is essential to provide the nucleophile for the first step of this mechanism.

Mechanism

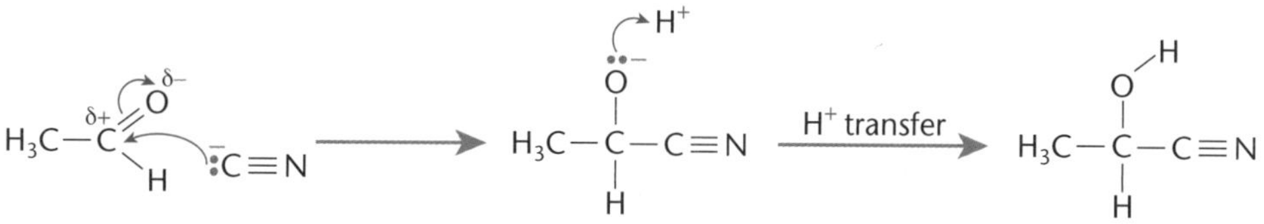

:CN⁻ nucleophile donates electron pair

'hydroxynitrile'

SALTERS M4

Increasing the carbon chain length

The nucleophilic addition of hydrogen cyanide is useful for increasing the length of a carbon chain. The nitrile product can then easily be reacted further in organic synthesis.

- Nitriles are easily ***hydrolysed*** by water in hot dilute acid to form a carboxylic acid.
- Nitriles are easily ***reduced*** by sodium in ethanol to form an amine.

The diagrams below show how 2-hydroxypropanenitrile, synthesised on page 95, can be converted into a carboxylic acid and an amine.

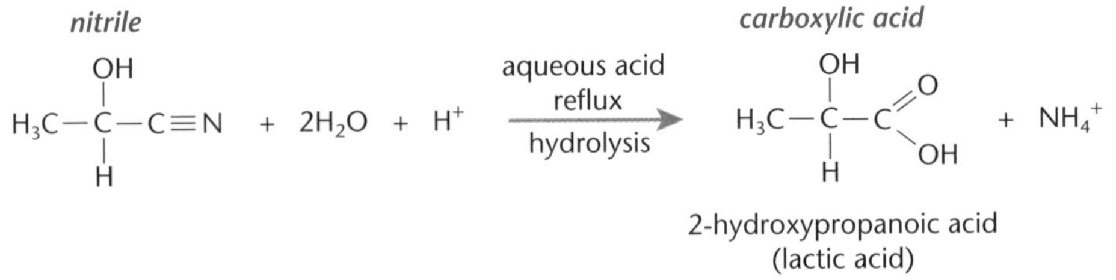

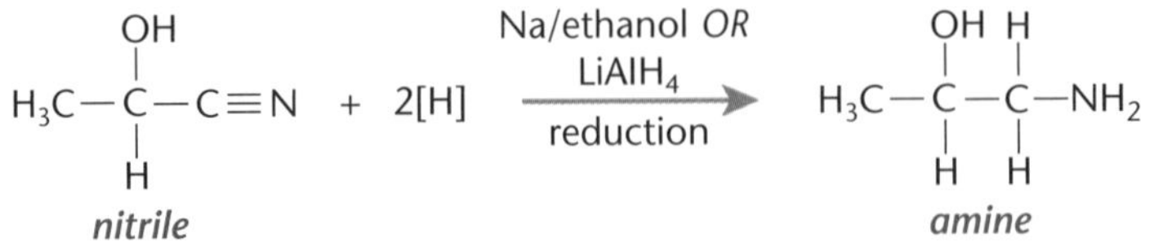

Testing for the carbonyl group

OCR M4

Carbonyl compounds produce an orange–yellow crystalline solid with Brady's reagent.

The carbonyl group can also be detected using IR spectroscopy, see page 143 (see also AS pages 119–121).

The carbonyl group can be detected using **Brady's reagent** – a solution of 2,4-dinitrophenylhydrazine (2,4-DNPH) in dilute acid.

- With 2,4-DNPH, both aldehydes and ketones produce bright **orange–yellow crystals** which identify the carbonyl group, C=O.

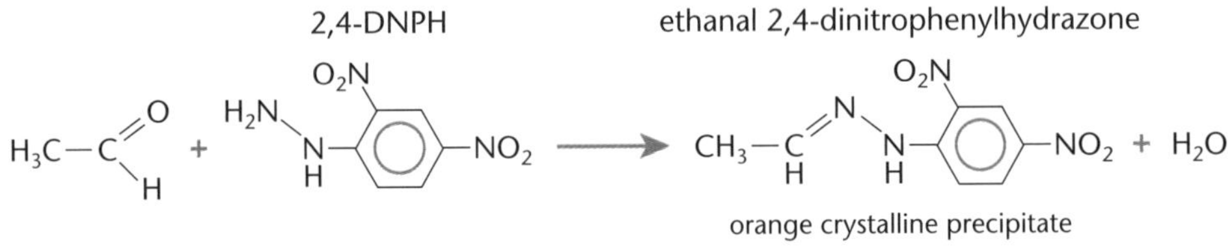

This is a **condensation reaction** – water is lost.

- If first recrystallised, the crystals have very sharp melting points which can be compared with known melting points from databases. Thus, the actual carbonyl compound can be identified.

Testing for the aldehyde group

OCR M4

An aldehyde can be distinguished from a ketone by using a combination of these two tests.

Both aldehyde and ketone produce a yellow–orange precipitate with 2,4-DNPH.

Only the aldehydes react in the three tests below.

Ketones cannot normally be oxidised and they **do not** react with Tollens' reagent, Fehling's solution or acidified dichromate(VI).

The oxidation of aldehydes provides the basis of chemical tests used to identify this functional group.

KEY POINT

In each test:
- the aldehyde **reduces** the reagents used in the test, producing a visible colour change
- the aldehyde is **oxidised** to a carboxylic acid.

$$RCHO + [O] \longrightarrow RCOOH$$

Heat aldehyde with Tollens' reagent

Tollens' reagent is a solution of silver nitrate in aqueous ammonia. The silver ions are reduced by an aldehyde producing a **silver mirror**.

Tollens' reagent contains the $[Ag(NH_3)_2]$ complex ion (see page 71).

> **KEY POINT**
> Tollens' reagent ⟶ **silver mirror**
> $Ag^+(aq) + e^- \longrightarrow Ag(s)$

Heat aldehyde with Fehling's (or Benedict's) solution

Fehling's or Benedict's solutions contain Cu^{2+} ions dissolved in aqueous alkali. The Cu^{2+} ions are reduced to copper(I) ions, Cu^+, producing a **brick-red precipitate** of Cu_2O.

Key points from AS

- **Oxidation of primary alcohols**
  Revise AS pages 113–114

> **KEY POINT**
> Benedict's or Fehling's solution ⟶ **brick-red precipitate** of Cu_2O
> $2Cu^{2+}(aq) + 2e^- + 2OH^-(aq) \longrightarrow Cu_2O(s) + H_2O(l)$

Heat aldehyde with $H^+/Cr_2O_7^{2-}$

Acidified dichromate(VI) can also be used to oxidise alcohols.

Concentrated sulfuric acid, H_2SO_4, is used as a source of H^+ ions and potassium dichromate(VI), $K_2Cr_2O_7$, as a source of $Cr_2O_7^{2-}$ ions. The **orange** $Cr_2O_7^{2-}$ ions are reduced to **green** chromium(III) ions, Cr^{3+}.

Progress check

1 (a) Draw the structure of:
   (i) 3-methylpentan-2-one
   (ii) 3-ethyl-2-methylpentanal.

2 **Three** isomers of C_4H_8O are carbonyl compounds.
   (a) (i) Show the formula for each isomer.
   (ii) Classify each isomer as an *aldehyde* or a *ketone*.
   (iii) Name each isomer.
   (b) Write the structural formulae of the **three** alcohols that could be formed by reduction of these isomers.

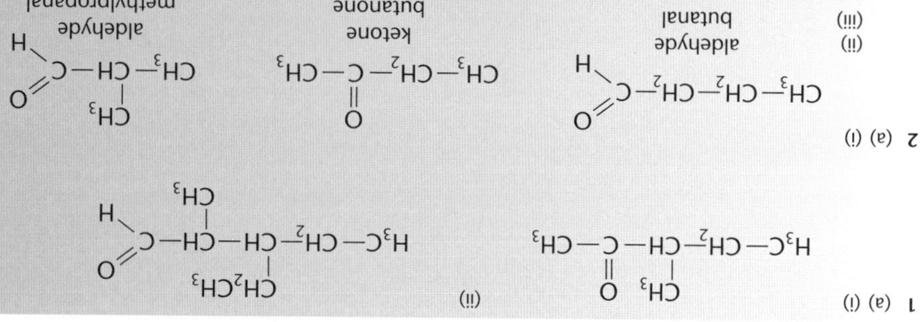

(b) $CH_3CH_2CH_2CH_2OH$; $CH_3CH_2CHOHCH_3$; $(CH_3)_2CHCH_2OH$

4.3 Carboxylic acids

After studying this section you should be able to:

- *understand the polarity and physical properties of carboxylic acids*
- *describe acid reactions of carboxylic acids to form salts*
- *describe the esterification of carboxylic acids with alcohols*

LEARNING SUMMARY

Carboxylic acids

General formula: $C_nH_{2n+1}COOH$ RCOOH

OCR	M4
SALTERS	M4

The carboxyl group

The functional group in carboxylic acids is the **carboxyl** group, COOH. Although this combines both the **carbon**yl group and a hydroxyl group its properties are very different from either.

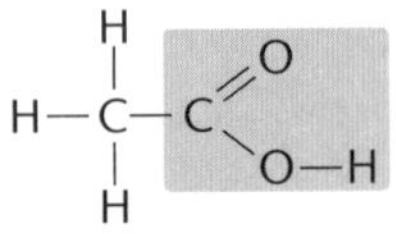

The combination of carbonyl and hydroxyl groups in the carboxyl group modifies the chemistry of both groups.

Carboxylic acids have their own set of reactions and react differently from carbonyl compounds and alcohols.

Naming of carboxylic acids

Numbering starts from the carbon atom of the carboxyl group.

4-methylpentanoic acid
$(CH_3)_2CH_2CH_2CH_2COOH$

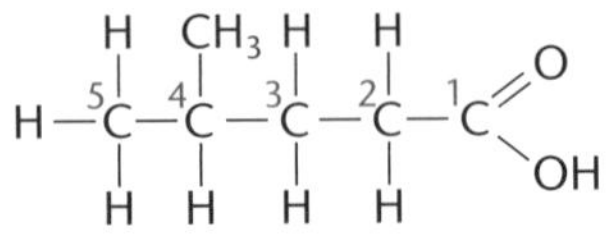

Natural carboxylic acids

Carboxylic acids are found commonly in nature. Their acidity is comparatively weak and their presence in food often gives a sour taste. Some examples of natural carboxylic acids are shown below.

structure	*name*	*natural source*
HCOOH	methanoic acid (*formic acid*)	ants, stinging nettles
CH_3COOH	ethanoic acid (*acetic acid*)	vinegar
COOH | COOH	ethanedioic acid (*oxalic acid*)	rhubarb
OH | H_3C—CH—COOH	2-hydroxypropanoic acid (*lactic acid*)	sour milk
CH_2—COOH | HO—C—COOH | CH_2—COOH	2-hydroxypropane-1,2,3-tricarboxylic acid (*citric acid*)	oranges, lemons

Polarity

The carboxyl group is a combination of two polar groups: the hydroxyl –OH, **and** carbonyl C=O groups. This makes a carboxylic acid molecule more polar than a molecule of an alcohol or a carbonyl compound.

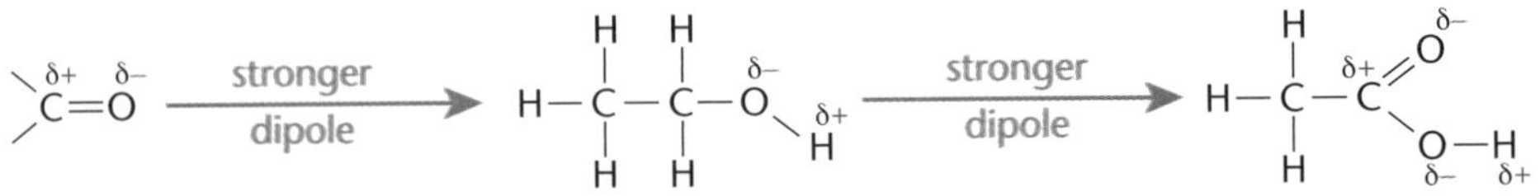

carbonyl group in aldehyde or ketone

hydroxyl group in alcohols

carboxyl group in carboxylic acids

The properties of carboxylic acids are dominated by the carboxyl group, COOH.

Physical properties of carboxylic acids

The COOH group dominates the physical properties of short-chain carboxylic acids. Hydrogen bonding takes place between carboxylic acid molecules, resulting in:

- higher melting and boiling points than alkanes of comparable M_r
- solubility in water.

The solubility of alcohols in water decreases with increasing carbon chain length as the non-polar contribution to the molecule becomes more important.

Carboxylic acids as 'acids'

OCR M4
SALTERS M4

Carboxylic acids are only weak acids because they only partially dissociate in water.

$$CH_3COOH \rightleftharpoons CH_3COO^- + H^+$$

- Only 1 molecule in about 100 actually dissociates.
- Only a small proportion of the potential H^+ ions is released.

Carboxylic acids are the 'organic acids'.

Carboxylates: salts of carboxylic acids

For more details of the dissociation of weak acids, see pages 29–30.

Carboxylic acid salts, 'carboxylates', are formed by neutralisation of a carboxylic acid by an alkali. In the example below, ethanoic acid produces ethanoate ions.

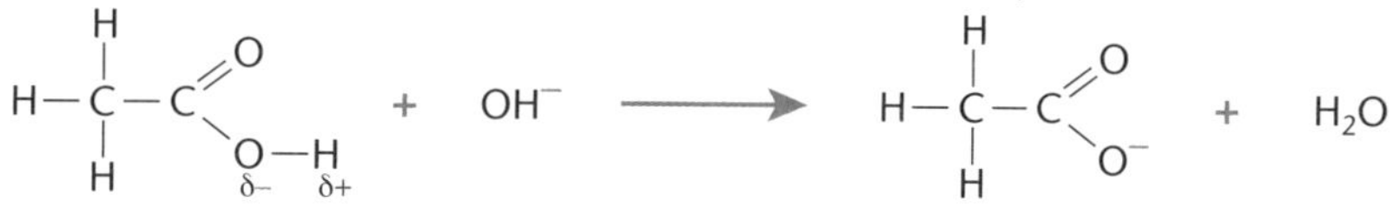

carboxylic acids exist in acidic conditions

carboxylates exist in alkaline conditions

On evaporation of water, a carboxylate salt crystallises out as an ionic compound. With aqueous sodium hydroxide as the alkali, sodium ethanoate, $CH_3COO^-Na^+$, is produced as the ionic salt.

Acid reactions of carboxylic acids

Key points from AS

- **Typical reactions of an acid**
  Revise AS page 39

Carboxylic acids take part in typical acid reactions. Note in the examples below that each salt formed is a carboxylate.

- They are **neutralised by alkalis**, forming a salt and water only.
  $CH_3COOH + NaOH \longrightarrow CH_3COONa + H_2O$
- They **react with carbonates**, forming a salt, carbon dioxide and water.
  $2CH_3COOH + CaCO_3 \longrightarrow (CH_3COO)_2Ca + CO_2 + H_2O$
- They **react with** reactive **metals**, forming a salt and hydrogen.
  $2CH_3COOH + Mg \longrightarrow (CH_3COO)_2Mg + H_2$

Carboxylic acids react by the usual 'acid reactions' producing carboxylate salts.

> Carboxylic acids are the only common organic group able to release carbon dioxide gas from carbonates. This provides a useful test to show the presence of a carboxyl group.
>
> **KEY POINT**

Properties of carboxylates

Carboxylates such as sodium ethanoate, $CH_3COO^-Na^+$, are ionic compounds. They have typical properties of an ionic compound.

- They are solids at room temperature with high melting and boiling points.
- They have a giant ionic lattice structure.
- They dissolve in water, totally dissociating into ions.

Esterification

OCR M4
SALTERS M4

Esters are formed by reaction of a carboxylic acid with an alcohol.

Esterification is the formation of an ester by reaction of a **carboxylic acid** with an **alcohol** in the presence of an **acid catalyst** (e.g. concentrated sulfuric acid).

carboxylic acid + alcohol ⟶ ester + water

The esterification of ethanoic acid by methanol is shown below:

$$CH_3COOH + CH_3OH \longrightarrow CH_3COOCH_3 + H_2O$$

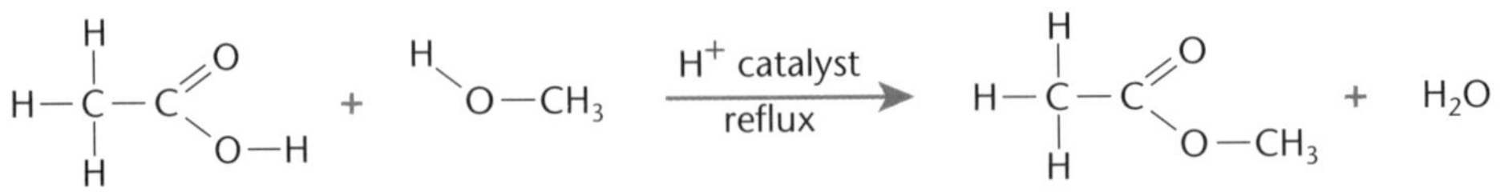

Conditions

- An acid catalyst (a few drops of conc. H_2SO_4) and reflux.
- The yield is usually poor due to incomplete reaction.

SALTERS M4

Purifying solids

The solvent is critical. Without a difference in solubility at high and low temperatures, hardly any crystals will appear on cooling.

Many organic compounds are solids and they will need to be purified. The easiest method of doing this is by **recrystallisation**.

The organic solid must be:

- very **soluble** in the solvent at **higher temperatures** – the solvent is saturated by the substance at higher temperatures
- **much less soluble**, or nearly so, in the solvent at **lower temperatures** – on cooling, the substance crystallises out, leaving impurities in solution.

Progress check

1 Write down the structural formula of:
(a) propanoic acid
(b) the propanoate ion.

2 Explain why a carboxylic acid has a higher boiling point than the corresponding alcohol.

1 (a) CH_3CH_2COOH
(b) $CH_3CH_2COO^-$

2 Carboxylic acid has both polar carbonyl and hydroxyl groups. Alcohols have hydroxyl group only. Therefore, a carboxylic acid has greater intermolecular forces.

4.4 Esters

After studying this section you should be able to:

LEARNING SUMMARY

- *understand the physical properties of esters*
- *describe the acid and base hydrolysis of esters*
- *understand that fats and oils are saturated and unsaturated esters*
- *describe the hydrolysis of fats and oils in soap making*

Esters

General formula: $C_nH_{2n+1}COOC_mH_{2m+1}$ RCOOR'

OCR M4
SALTERS M4

The functional group in esters is the COOR' group, with an alkyl group in place of the acidic proton of a carboxylic acid.

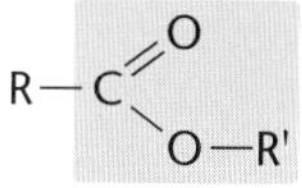

Naming of esters

The name of an ester is based on the carboxylic acid from which the ester is derived. The ester below, methyl butanoate, is derived from butanoic acid C_3H_7COOH with a methyl group in place of the acidic proton.

butanoate from butanoic acid $CH_3CH_2CH_2COOH$

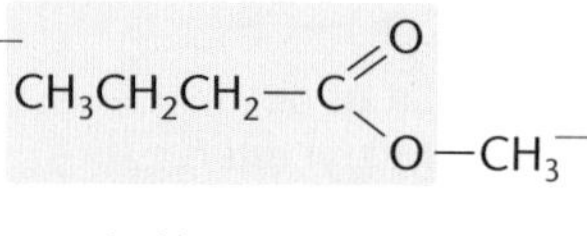

methyl from methanol CH_3OH

methyl butanoate

In the name of **methyl butanoate**:

- the alkyl group comes first as the ...*yl*: **methyl**
- the carboxylic acid part comes second as the ...*oate*: **butanoate**

Natural esters

Esters are found commonly in nature as fats and oils. They often have pleasant smells and contribute to the flavouring of many foods as shown below.

Esters are common organic compounds present in fats and oils. They are also used as solvents, plasticisers, in food flavourings and as perfumes.

structure	*name*	*source*
$HCOOCH_3$	methyl methanoate	raspberries
$C_3H_7COOC_4H_9$	butyl butanoate	pineapple

Physical properties of esters

Unlike carboxylic acids, esters are neutral. They are less polar than carboxylic acids. The absence of an –OH group means that esters cannot form hydrogen bonds and are generally insoluble in water.

Hydrolysis of esters

OCR M4
SALTERS M4

The hydrolysis of an ester is the reverse reaction to esterification (see page 100):

$$\text{ester} + \text{water} \underset{\textit{esterification}}{\overset{\textit{hydrolysis}}{\rightleftharpoons}} \text{carboxylic acid} + \text{alcohol}$$

Note that this reaction is reversible. The direction of reaction can be controlled by the reagents and reaction conditions used.

KEY POINT

Hydrolysis is the breaking down of a compound using **water** as the reagent.

> Esterification **produces** water.
> Hydrolysis **reacts** with water.

Hydrolysis takes place by refluxing the ester with dilute aqueous acid or alkali.

- Acid hydrolysis ⟶ alcohol + carboxylic acid.
- Alkaline hydrolysis ⟶ alcohol + carboxylate.

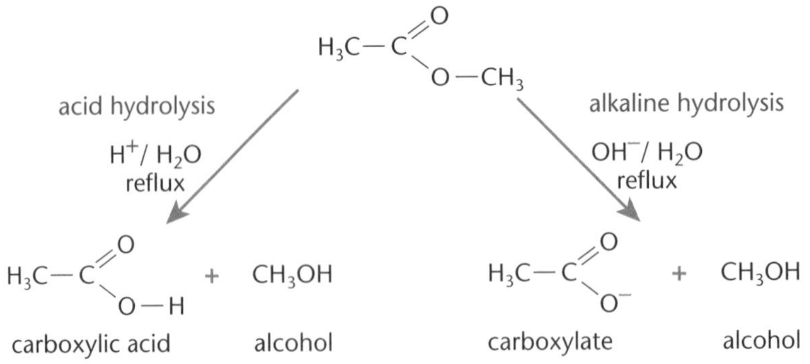

Fats and oils

OCR M4
SALTERS M5

> Most fats and oils are mixed esters from different fatty acids. So 'R' in these structures may refer to three different chains.

Natural fats and oils are *triglyceryl esters* of fatty acids (long chain carboxylic acids) and propane-1,2,3-triol (*glycerol*).

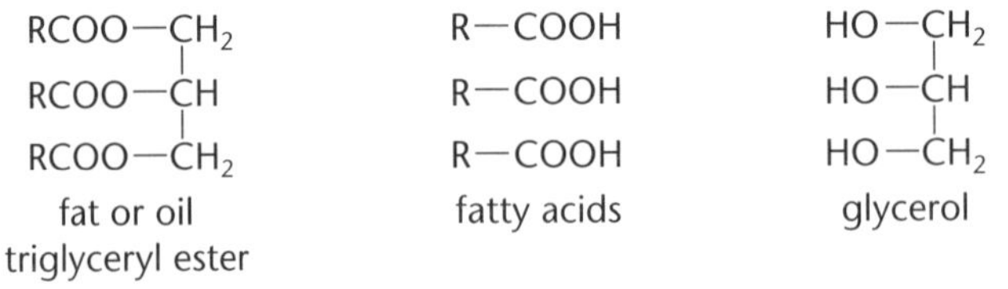

Saturated and unsaturated fats

Saturated fats tend to be solids at room temperature and are mainly from animal products. Unsaturated fats tend to be oils and are mainly from vegetable products.

Saturated and unsaturated fatty acids can be obtained from fats and oils by hydrolysis (see page 103). An unsaturated fatty acid would have at least one double C=C bond in one of the alkyl carbon chains.

Examples of saturated and unsaturated fatty acids

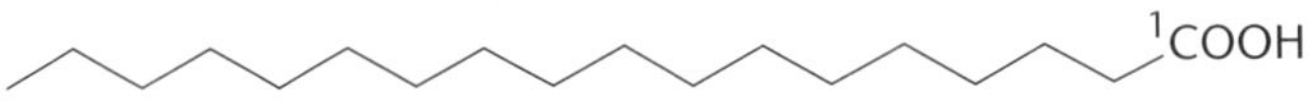

octadecanoic acid, 18,0

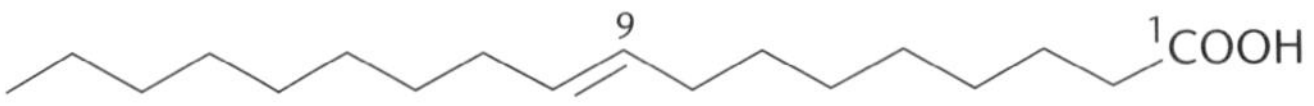

octadec-9-enoic acid, 18,1(9);

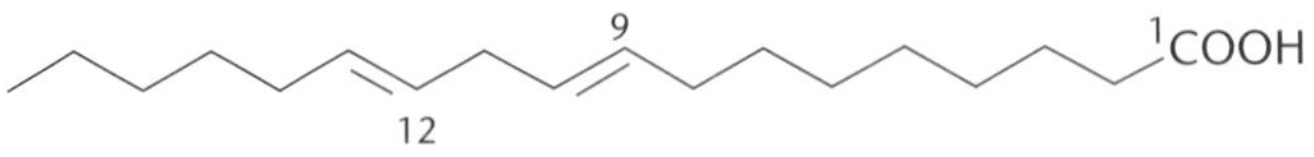

octadeca-9,12-dienoic acid, 18,2(9,12)

- Note that the unsaturated fatty acids above are *trans* acids. The *cis* isomers are less compact and are far healthier.
- Many triglyceryl esters contain different alkyl carbon chains derived from different fatty acids.

> Scientists have discovered that *trans* fatty acids pose health problems, increasing cholesterol with an increased risk of coronary heart disease, strokes and obesity.

In margarine production, vegetable oils are hardened by partial **catalytic hydrogenation**.

- Partial hydrogenation leaves some double bonds intact. The oil is reacted just enough to solidify the oily texture. This prevents the formation of **polysaturates**, with no C=C double bonds, which are linked to unhealthy diets.
- Margarines are often sold as partially hydrogenated **polyunsaturates** to promote the health value of unsaturated fats.

Biodiesel

OCR M4

Transesterification is also being investigated for the development of low-fat spreads as an alternative to hydrogenation of vegetable oils to produce margarine.

Esters of fatty acids are being developed for use as fuels such as biodiesel. Use of biodiesel as a fuel increases the contribution to energy requirements from renewable fuels. Biodiesel is a mixture of esters of long chain carboxylic acids.

Vegetable oils can be converted into biodiesel by **transesterification** using methanol in the presence of an alkaline catalyst, e.g.

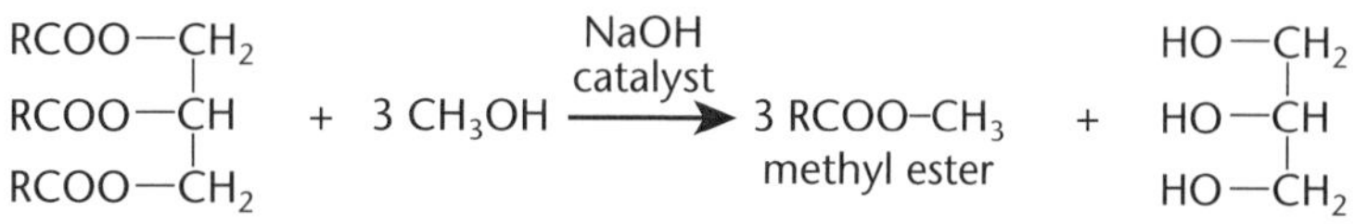

The methyl ester is the easiest to prepare but methanol has the disadvantage that it is toxic. Research is being done to develop the ethyl ester made from ethanol. Ethanol is easy to produce from fermentation of sugars.

Progress check

1 How is methyl propanoate made from a named carboxylic acid and alcohol?

2 (a) Explain what is meant by the term *hydrolysis*.
(b) Name the products of the acid and alkaline hydrolysis of ethyl butanoate.

1 Reflux propanoic acid with methanol in the presence of a few drops of concentrated sulfuric acid as catalyst.

2 (a) Breaking down a molecule using water.
(b) Acid hydrolysis: ethanol and butanoic acid.
Alkaline hydrolysis: ethanol and butanoate ions.

4.5 Acylation

After studying this section you should be able to:

- *describe the formation of acyl chlorides from carboxylic acids*
- *describe nucleophilic addition-elimination reactions of acyl chlorides*
- *know the advantages of ethanoic anhydride for industrial acylations*

LEARNING SUMMARY

Acyl chlorides

General formula: $C_nH_{2n+1}COCl$ RCOCl

SALTERS M4

The functional group in acyl chlorides is the COCl group:

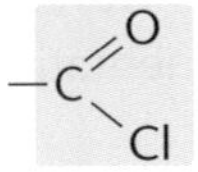

Acyl chlorides are named from the parent carboxylic acid.

The examples below show that the suffix '*-oic acid*' is changed to '*-oyl chloride*' in the corresponding acyl chloride.

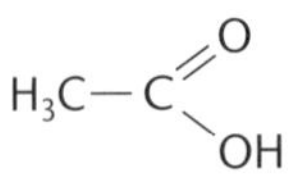
ethanoic acid

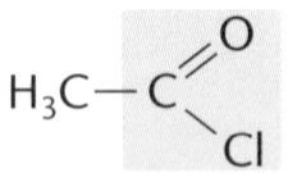
ethanoyl chloride

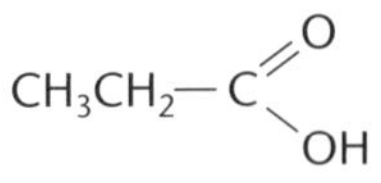
propanoic acid

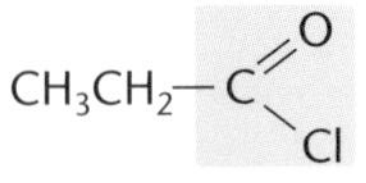
propanoyl chloride

Properties

Acyl chlorides are the most reactive organic compounds that are commonly used. Unlike most other organic functional groups, acyl chlorides do **not** occur naturally because of their high reactivity with water.

Acyl chlorides are the only common organic compounds that react violently with water.

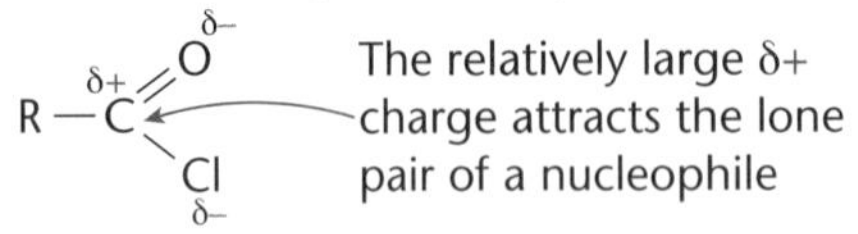

The electron-withdrawing effects of both the **chlorine** atom and carbonyl **oxygen** atom produces a relatively large δ+ charge on the carbonyl carbon atom. Nucleophiles are strongly attracted to the electron-deficient carbon atom, increasing the reactivity.

Reactions of acyl chlorides with nucleophiles

SALTERS M4

Nucleophiles of the type H–Y**:** react readily with acyl chlorides.

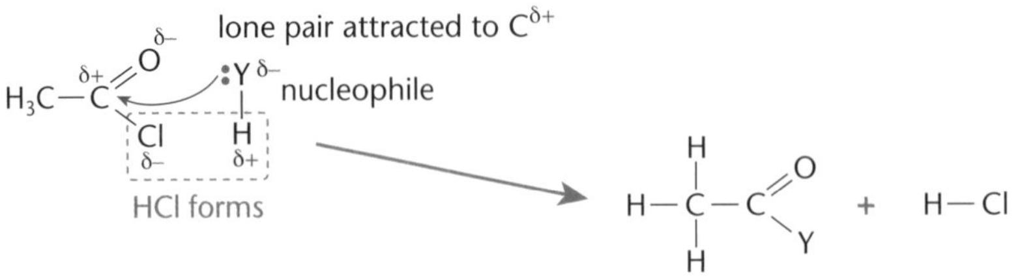

This is an **addition-elimination** reaction involving:

- addition of HY across the C=O double bond followed by
- elimination of HCl.

Addition-elimination reactions of acyl chlorides

Acyl chlorides can be reacted with different nucleophiles to produce a range of related functional groups. In each addition-elimination reaction, hydrogen chloride is eliminated as the second product.

The reaction scheme below shows the reactions of an acyl chloride RCOCl with the nucleophiles water, H_2O, methanol, CH_3OH, ammonia, NH_3 and methylamine, CH_3NH_2.

The high reactivity of an acyl chloride means that these reactions take place at room temperature.

Acyl chlorides also react with phenols to form esters.

carboxylic acid **ester**

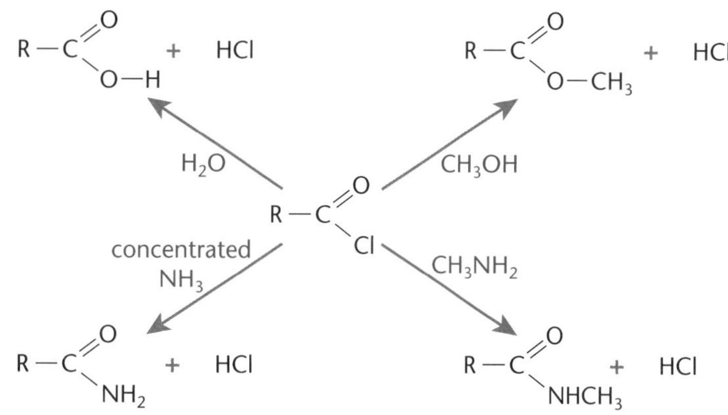

primary amide **secondary amide**

Ammonia and methylamine are bases – they react with the HCl eliminated:

$$NH_3 + HCl \longrightarrow NH_4^+Cl^-$$
$$CH_3NH_2 + HCl \longrightarrow CH_3NH_3^+Cl^-$$

Acid anhydrides

OCR M4

Acyl chlorides are ideal for small-scale preparations in the laboratory. However, they are too expensive and too reactive for large-scale preparations, for which they are replaced by acid anhydrides.

The equation below shows the reaction between ethanoic anhydride and methanol:

Notice the molecular structure of ethanoic anhydride – two ethanoic acid molecules have been condensed together with loss of H_2O. Hence the 'anhydride' from ethanoic acid.

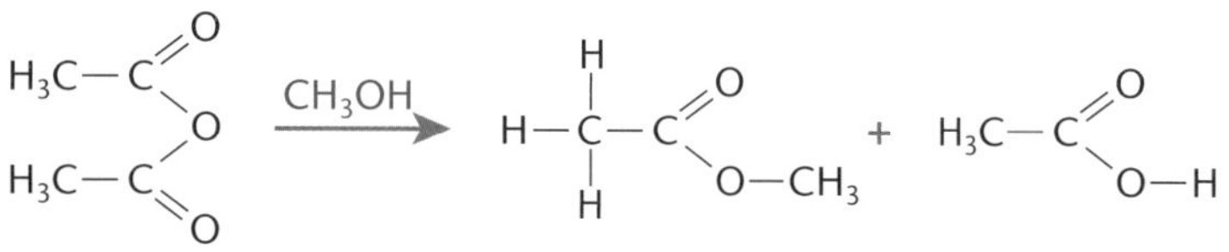

acid anhydride

carboxylic acid as second product

- The reaction is slower and easier to control on a large scale.
- This is the same essential reaction as with an acyl chloride but RCOOH forms instead of HCl as the second product.

Synthesis of aspirin

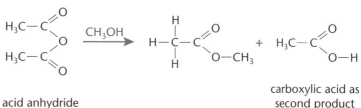

Aspirin can be synthesised by acylation of 2-hydroxybenzenecarboxylic acid (*salicylic acid*) using ethanoic anhydride.

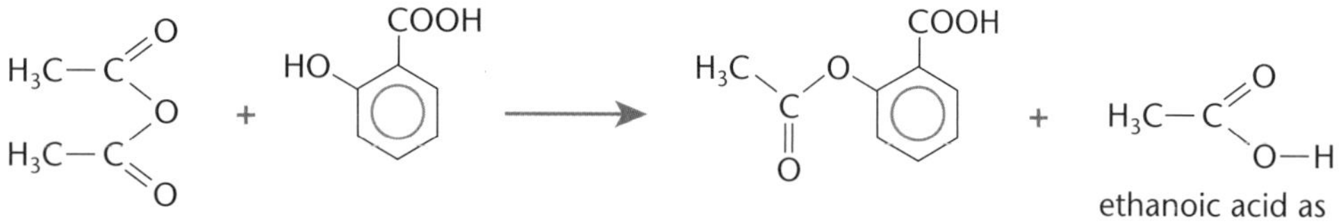

aspirin

ethanoic acid as second product

Progress check

1 Write down the structural formula of the organic product formed from the reaction of propanoyl chloride with:
(a) water
(b) ammonia.

2 What could you react together to make N-ethylbutanamide, $CH_3CH_2CH_2CONHCH_2CH_3$?

3 Write an equation for the preparation of aspirin from 2-hydroxybenzoic acid using an acyl chloride.

1 (a) CH_3CH_2COOH
(b) $CH_3CH_2CONH_2$

2 $CH_3CH_2CH_2COCl$ and $CH_3CH_2NH_2$

3

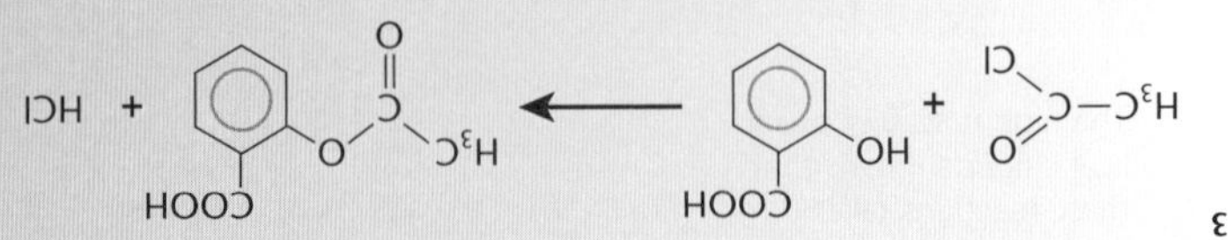

4.6 Arenes

After studying this section you should be able to:

- *apply rules for naming simple aromatic compounds*
- *understand the delocalised model of benzene*
- *explain the resistance to addition of benzene compared with alkenes*

LEARNING SUMMARY

Aromatic organic compounds

OCR M4
SALTERS M5

Organic compounds with pleasant smells were originally classified as aromatic compounds. Many of these contain a benzene ring in their structure and nowadays an aromatic compound is one structurally derived from benzene, C_6H_6.

The diagrams below show different representations of a benzene molecule. It is usual practice to omit the carbon and hydrogen labels.

Arenes burn with a smoky flame. This reflects the relatively low hydrogen to carbon ratio compared to alkanes and alkenes.

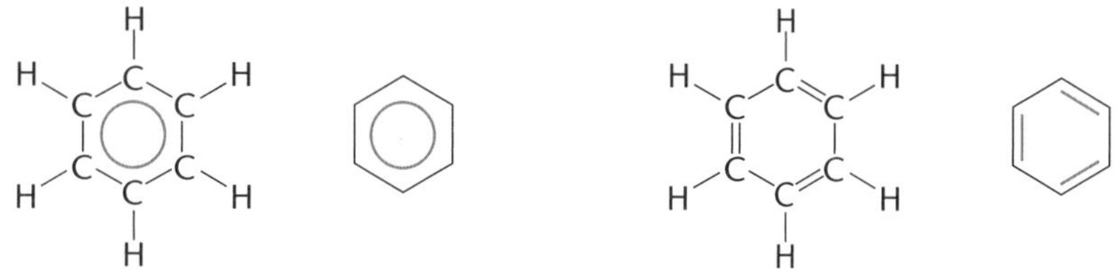

Arenes

An **arene** is an aromatic hydrocarbon containing a benzene ring. Benzene is the simplest arene and substituted arenes have alkyl groups attached to the benzene ring. Examples of arenes are shown below.

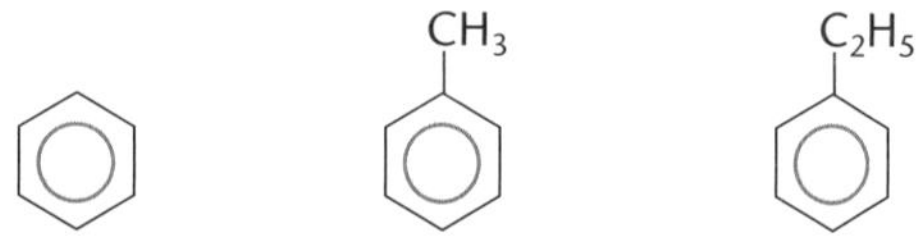

benzene methylbenzene ethylbenzene

Functional groups

Arenes can have a functional group next to an **aryl** group (a group containing a benzene ring):

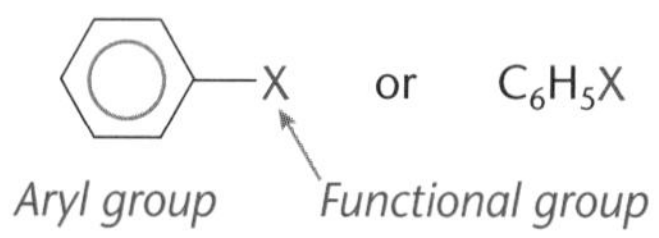

An aryl group contains a benzene ring.

- An aryl group is often represented simply as Ar—.
- The simplest aryl group is the **phenyl** group, C_6H_5, derived from benzene, C_6H_6.

Naming of aromatic organic compounds

The benzene ring of an aromatic compound is numbered from the carbon atom attached to a functional group or alkyl side-chain.
The names can be derived in two ways:

- Some compounds (e.g. hydrocarbons, chloroarenes and nitroarenes) are regarded as substituted benzene rings.

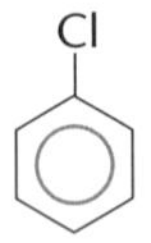
chlorobenzene

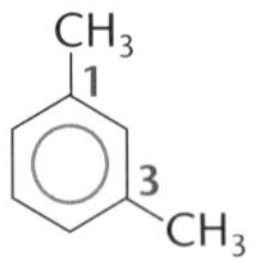
1,3-dimethylbenzene

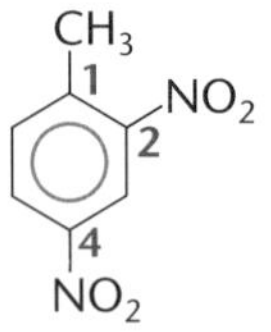
2,4-dinitromethylbenzene

You should be able to suggest names for simple arenes.

- Other compounds (e.g. phenols and amines) are considered as phenyl compounds of a functional group.

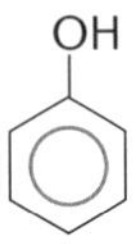
phenol
not hydroxybenzene

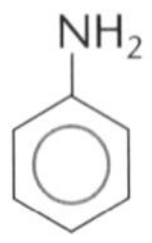
phenylamine
not aminobenzene

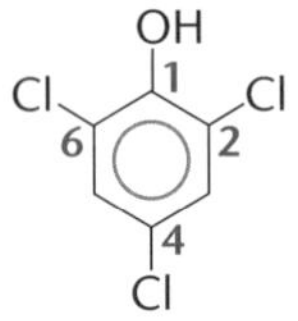
2,4,6-trichlorophenol

The stability of benzene

OCR M4
SALTERS M5

Two structures are used to represent benzene:

- the **Kekulé** structure, developed between 1865 and 1872
- the modern **delocalised**, structure developed in the 1930s.

The Kekulé structure of benzene

The Kekulé model of a benzene molecule has alternate double and single bonds making up the ring.

The chemical name for the Kekulé structure of benzene is **cyclohexa-1,3,5-triene,** after the positions of the double bonds in the ring.

The original Kekulé structure showed benzene as a hexagonal molecule with alternate double and single bonds. Each carbon atom is attached to one hydrogen atom. This model was later modified to one with two isomers, rapidly interconverting into one another.

The Kekulé structure of benzene

The delocalised structure of benzene

The delocalised structure of benzene shows a benzene molecule as a hybrid state between Kekulé's two isomers with no separate single and double bonds.

The delocalised model of a benzene molecule has identical carbon–carbon bonds making up the ring.

In this hybrid state:

- each carbon atom contributes one electron from its p-orbital to form π-bonds
- the π-bonds are spread out or **delocalised** over the whole ring.

The double bond in an alkene is a **localised** π-bond between two carbon atoms. In the delocalised structure of benzene, each bond is identical, with electron density spread out to encompass and stabilise the whole ring. The diagram below shows one of the π-bonds formed by delocalisation of electrons in a benzene molecule.

Key point from AS

- **Alkenes**
  Revise AS page 107

This model helps to explain the low reactivity of benzene compared with alkenes.

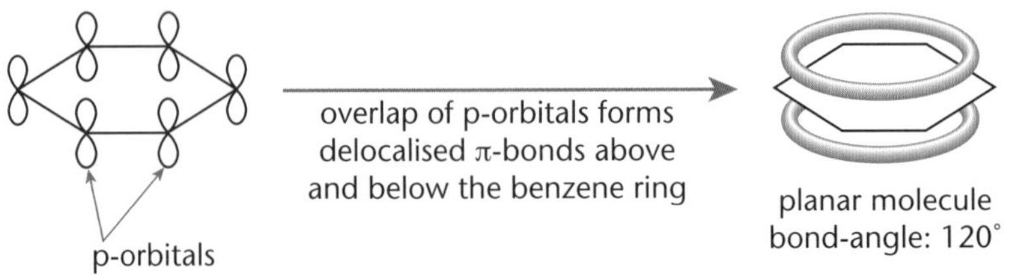

Although the Kekulé structure is used for some purposes, the delocalised structure is a better representation of benzene. You will find both representations in books.

The delocalised structure of a benzene molecule is shown as a hexagon to represent the carbon skeleton and a circle to represent the six delocalised electrons.

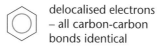

delocalised electrons
– all carbon-carbon
bonds identical

The shape of a benzene molecule

Key points from AS

- **Electron-pair repulsion theory**
  Revise AS pages 50–51

Using electron-pair repulsion theory:

- there are **three** centres of electron density surrounding each carbon atom
- the shape around each carbon atom is trigonal planar with bond angles of 120°.

This results in a benzene molecule that is **planar**.

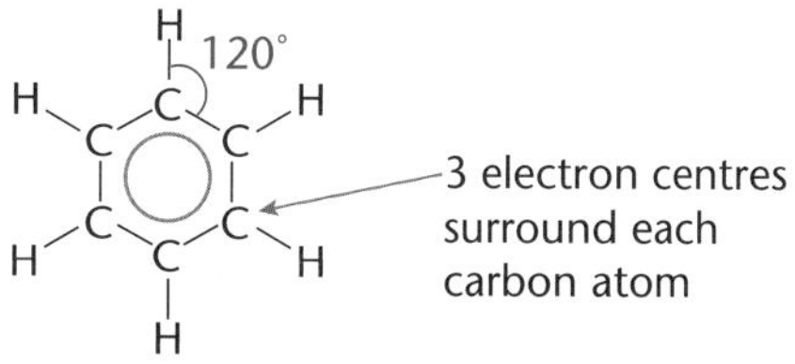

Experimental evidence for delocalisation

OCR M4
SALTERS M5

Bond length data

The Kekulé structure of benzene as cyclohexa-1,3,5-triene suggests two bond lengths for the separate single and double bonds:

- C—C bond length = 0.154 nm
- C=C bond length = 0.134 nm

Experiment shows only one C—C bond length of 0.139 nm, between the bond lengths for single and double carbon-carbon bonds.

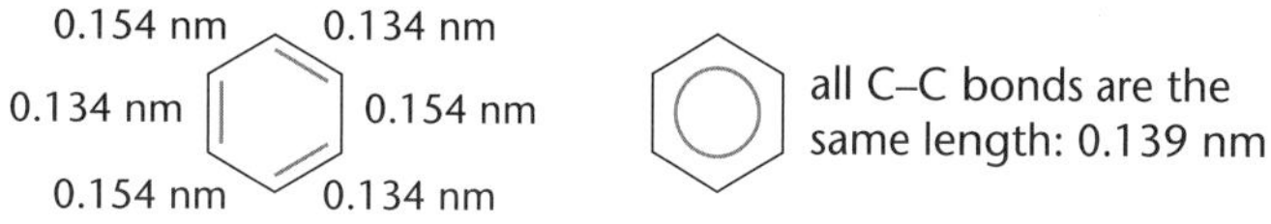

KEY POINT

This shows that each carbon-carbon bond in the benzene ring is intermediate between a single and a double bond.

Thermochemical evidence

Hydrogenation of cyclohexene

Each molecule of cyclohexene has **one** C=C double bond. The enthalpy change for the reaction of cyclohexene with hydrogen is shown below:

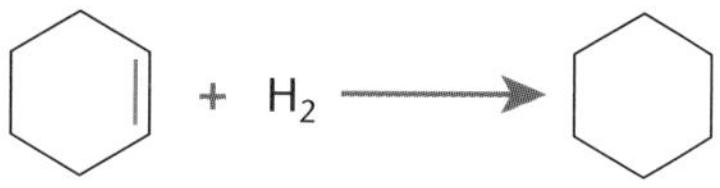

$\Delta H^{\ominus} = -120\ \text{kJ mol}^{-1}$

Hydrogenation of benzene

The stability of benzene can also be demonstrated using thermochemical data for the reactions of bromine with cyclohexene and benzene.

The Kekulé structure of benzene as cyclohexa-1,3,5-triene has **three** double C=C bonds. It would be expected that the enthalpy change for the hydrogenation of this structure would be three times the enthalpy change for the **one** C=C bond in cyclohexene.

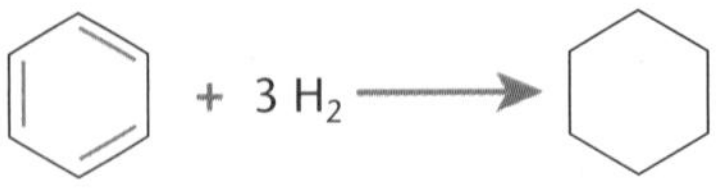

predicted enthalpy change:
$\Delta H^\ominus = 3 \times -120 = -360$ kJ mol^{-1}

- When benzene is reacted with hydrogen, the enthalpy change obtained is far less exothermic, $\Delta H^\ominus = -208$ kJ mol^{-1}.

KEY POINT

- The difference between the thermochemical data for cyclohexa-1,3,5-triene and benzene suggests that benzene has **more stable bonding** than the Kekulé structure.
- The delocalisation or resonance energy of benzene of –152 kJ mol^{-1} is the difference between the two enthalpy changes above. This is extra energy that must be provided to break the delocalised benzene ring.

Resistance to reaction with electrophiles

Further evidence for delocalisation is provided by the relative resistance to bromination of benzene, compared with alkenes.

- The delocalised π-bonds in benzene have less electron density than the localised π-bonds of the C=C bond in alkenes.
- Electrophiles such as bromine experience less attraction from benzene's π-bonds than the π-bonds in an alkene.

For further details, see 'Comparing halogenation of arenes and alkenes' (page 112).

Progress check

1 Name the **three** aromatic isomers of $C_6H_4Br_2$.

2 State two pieces of experimental evidence that support the delocalised structure of benzene.

3 Explain why alkenes, such as cyclohexene, are so much more reactive with electrophiles than arenes such as benzene.

1 1,2-dibromobenzene; 1,3-dibromobenzene; 1,4-dibromobenzene.

2 The enthalpy change of hydrogenation of benzene is less exothermic than three times the enthalpy change of hydrogenation of cyclohexene.
All the carbon-carbon bonds in the benzene ring are the same length in between a double and single bond.

3 Alkenes have localised π-bonds with a larger electron density than the delocalised π-bonds in benzene. The greater electron density is able to attract electrophiles more strongly. Also, the stability of the benzene ring must be disrupted if benzene is to react.

4.7 Reactions of arenes

After studying this section you should be able to:

LEARNING SUMMARY

- *describe electrophilic substitution of arenes: nitration, halogenation, alkylation, acylation and sulfonation*
- *describe the mechanism of electrophilic substitution in arenes*
- *understand the importance of reactions of arenes in the synthesis of commercially important materials*

Electrophilic substitution reactions of arenes

OCR M4
SALTERS M5

Many electrophiles react with alkenes by **addition**. However, electrophiles react with arenes by **substitution**, replacing a hydrogen atom on the ring. The difference in behaviour results from the high stability of the **delocalised** π-bonds in benzene compared with the **localised** π-bonds in alkenes (see pages 108–109).

KEY POINT

- Benzene reacts with only very reactive electrophiles.
- The typical reaction of an arene is **electrophilic substitution**.

In this section, reactions of benzene are discussed to illustrate electrophilic substitution reactions of arenes.

Nitration of arenes

OCR M4
SALTERS M5

The nitration of arenes produces aromatic nitro compounds, which are important for the synthesis of many important products including explosives and dyes (see page 120).

Nitration of benzene

Benzene is nitrated by concentrated nitric acid at 55°C in the presence of concentrated sulfuric acid, which acts as a catalyst.

Although some heat is required (55°C), too much may give further nitration of the benzene ring forming 1,3-dinitrobenzene.

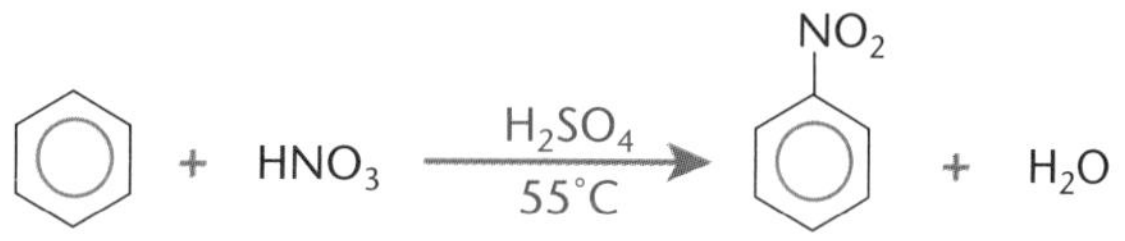

Mechanism

The nitronium ion is also called a **nitryl cation**.

- The role of the concentrated sulfuric acid is to generate the **nitronium ion**, NO_2^+, as the active electrophile:

$$HNO_3 + H_2SO_4 \longrightarrow H_2NO_3^+ + HSO_4^-$$
$$H_2NO_3^+ \longrightarrow NO_2^+ + H_2O$$

- The powerful NO_2^+ electrophile then reacts with benzene.

This is **electrophilic substitution.**

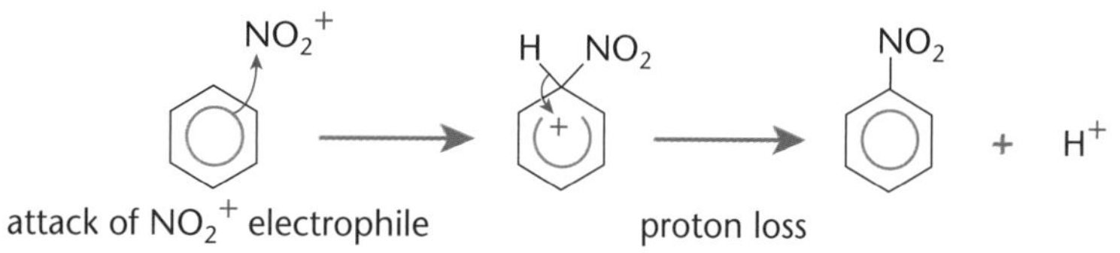

- The H^+ formed regenerates a molecule of H_2SO_4.

$$H^+ + HSO_4^- \longrightarrow H_2SO_4$$

The H_2SO_4 therefore acts as a **catalyst**.

- The H_2SO_4 molecule reacts with more nitric acid to form more nitronium ions.

Comparing halogenation of arenes and alkenes

OCR M4
SALTERS M5

Key points from AS

- **Alkenes**
  Revise AS page 107
- **Addition reactions of alkenes**
  Revise AS pages 108–110

Bromination of alkenes

Alkenes and cycloalkenes react with bromine by **electrophilic addition**. The reaction takes place in the dark at room temperature. The bromination of cyclohexene is shown below:

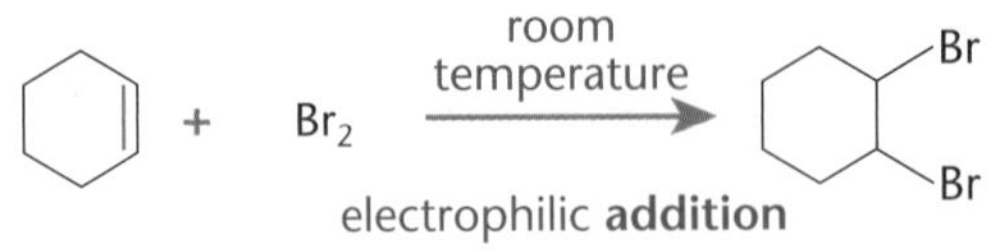

- Using the same conditions with benzene, there is **no reaction**.
- The stability of the delocalised system resists addition that would disrupt this stability (see pages 108–109).

Bromination of arenes

Arenes react with bromine by **electrophilic substitution**. The reaction takes place only in the presence of a **halogen carrier**, which acts as a catalyst.

Suitable halogen carriers include:

- iron
- aluminium halides, e.g. $AlCl_3$ for chlorination; $AlBr_3$ for bromination.

The bromination of benzene is shown below.

KEY POINT

Bromination of alkenes and arenes are different types of reaction:

alkenes: electrophilic **addition**

arenes: electrophilic **substitution**

OCR M4

Mechanism

- Iron first reacts with bromine forming iron(III) bromide, $FeBr_3$.
- $FeBr_3$ acts as a halogen carrier, polarising the Br—Br bond.

$$Br_2 + FeBr_3 \longrightarrow Br^{\delta+}\text{—}Br^{\delta-}\ FeBr_3$$

- This electrophile reacts with benzene.

This is a similar principle to the nitration of the benzene ring.

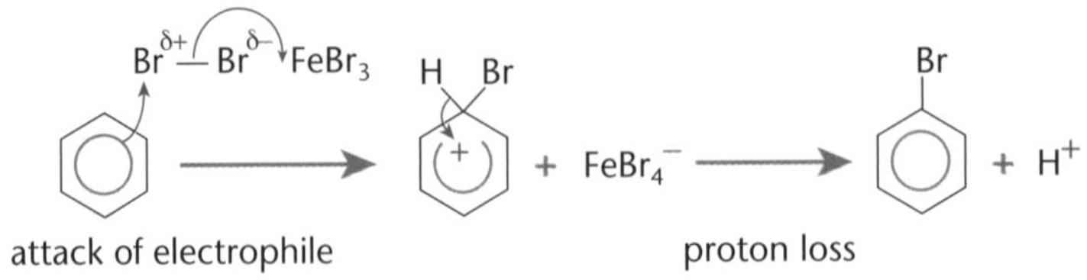

- The H^+ formed generates $FeBr_3$.

$$H^+ + FeBr_4^- \longrightarrow FeBr_3 + HBr$$

- $FeBr_3$ can now polarise more bromine molecules.

The $FeBr_3$ therefore acts as a **catalyst**.

Alkylation of arenes

SALTERS M5

Alkylation reactions of arenes are commonly known as *Friedel–Crafts* reactions. These are very important reactions industrially as they provide a means of introducing an alkyl group onto the benzene ring.

Alkylation of benzene

This is a similar principle to the halogenation of the benzene ring.

As with halogenation, a halogen carrier is required to generate a more reactive electrophile. With chloroethane, ethylbenzene is formed.

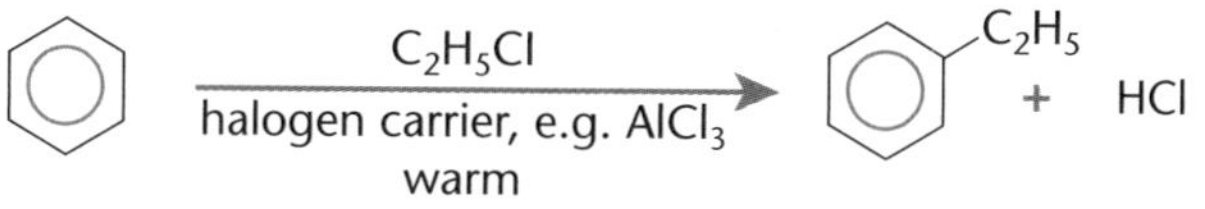

Acylation of arenes

SALTERS M5

Acylation introduces an acyl group such as $CH_3C{=}O$ onto the benzene ring. An acyl chloride is used in the presence of a halogen carrier.

The mechanism for acylation is similar to that of alkylation.

The acylium ion is then able to react with the benzene ring, introducing an acyl group onto the ring.

Acylation of benzene

The acylation of benzene is another example of a *Friedel–Crafts* reaction and is also important industrially.

Using ethanoyl chloride, CH_3COCl, the ethanoyl group $CH_3C{=}O$ can be introduced onto the benzene ring.

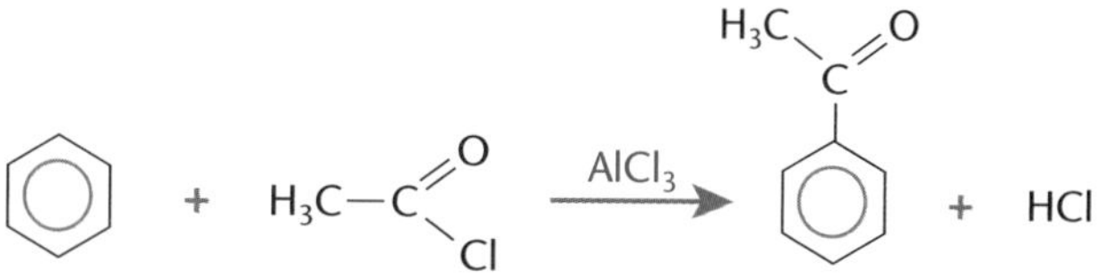

Sulfonation of arenes

SALTERS M5

Sulfonation of arenes produces sulfonic acids. Sodium salts of sulfonic acids are used for detergents and fabric conditioners.

Sulfonation of benzene

Benzene reacts with fuming sulfuric acid (concentrated sulfuric acid saturated with sulfur trioxide) forming a sulfonic acid.

The reaction takes place between benzene and sulfur trioxide.

benzenesulfonic acid

Progress check

1 What is the common type of reaction of arenes?

2 Benzene reacts with bromine, nitric acid and chloroethane.

(a) Using C_6H_6 to represent benzene and C_6H_5 to represent the phenyl group, write balanced equations for each of these reactions.

(b) State the essential conditions that are needed for each of these reactions.

1 Electrophilic substitution.

2 (a) $C_6H_6 + Br_2 \longrightarrow C_6H_5Br + HBr$

$C_6H_6 + HNO_3 \longrightarrow C_6H_5NO_2 + H_2O$

$C_6H_6 + C_2H_5Cl \longrightarrow C_6H_5C_2H_5 + HCl$

(b) With bromine, a halogen carrier such as Fe.

With nitric acid, concentrated sulfuric acid at 55°C.

With chloroethane, a halogen carrier such as Fe.

4.8 Phenols and alcohols

After studying this section you should be able to:

- *state the uses of phenols in antiseptics and disinfectants*
- *describe the reactions of phenol with sodium and bases*
- *explain the ease of bromination of phenol compared with benzene*

LEARNING SUMMARY

Phenols

General structure: ArOH (Ar is an aromatic ring)

OCR M4
SALTERS M4

Phenols and alcohols both have a hydroxyl functional group. In alcohols, the hydroxyl group is bonded to a carbon chain but phenols have an aromatic ring bonded directly to a hydroxyl (–OH) group. Some phenols are shown below.

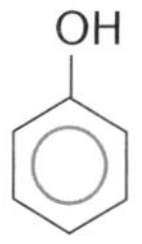

phenol, C_6H_5OH

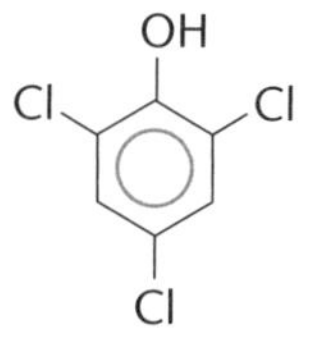

2,4,6-trichlorophenol (TCP)

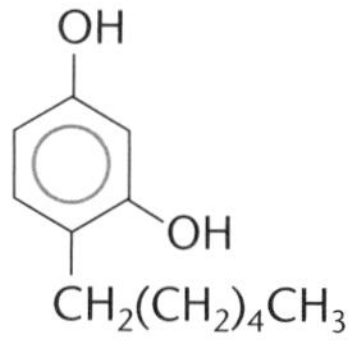

4-hexyl-3-hydroxyphenol (in cough drops)

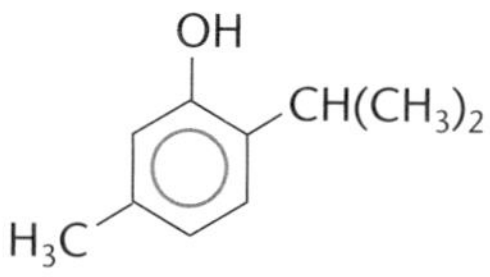

thymol (in thyme)

Uses

Dilute solutions of phenol are used as disinfectants and phenol was probably the first antiseptic. However, phenol is toxic and can cause burns to the skin. Less toxic phenols are used nowadays in antiseptics.

With $FeCl_3$ solutions, phenols produce a violet colour. This colour change is often used as a test for the phenol group.

Phenols are also used in the production of plastics, such as the widely used phenol-formaldehyde resins. More complex phenols, such as thymol shown above, can be used as flavourings and aromas and these are obtained from essential oils of plants.

Properties

Like alcohols, a phenol has a **hydroxyl group** which is able to form intermolecular **hydrogen bonds**. The resulting properties include:

- higher melting and boiling points than hydrocarbons with similar relative molecular masses
- some solubility in water.

Reactions of phenol as an acid

OCR M4
SALTERS M4

Unlike the neutrality shown by alcohols, phenols are very weak acids. The hydroxyl group donates a proton, H^+, by breaking the O–H bond.

$$C_6H_5OH \rightleftharpoons H^+ + C_6H_5O^-$$

Phenols take part in some typical acid reactions, reacting with sodium and bases to form ionic salts called **phenoxides**. However, the acidity is too weak to release carbon dioxide from carbonates and hydrogencarbonates (see carboxylic acids, page 99).

Reaction with sodium

As with the hydroxyl groups in water and ethanol, phenol reacts with sodium, releasing hydrogen gas and forming a solution of sodium phenoxide, C_6H_5ONa.

$$2C_6H_5OH + 2Na \longrightarrow 2C_6H_5ONa + H_2$$

Compare these reactions with those of carboxylic acids page 99.

Reaction with an alkali

Phenol is a weak acid and is neutralised by aqueous alkalis. For example, aqueous sodium hydroxide forms the salt sodium phenoxide. Alcohols and water are not acidic enough to neutralise alkalis.

$$C_6H_5OH + NaOH \longrightarrow C_6H_5ONa + H_2O$$

Electrophilic substitution of the aryl ring of phenol

OCR M4

The aryl ring of phenol has a greater electron density than benzene because the adjacent oxygen reinforces the electron density in the ring. A p-orbital on the oxygen donates an electron pair to the benzene ring.

This results in:

- a greater electron density in the aryl ring, activating the ring
- greater reactivity of the ring towards electrophiles.

Bromination of phenol

- Phenol reacts directly with bromine. However, Benzene reacts with bromine only in the presence of a halogen carrier.
- Phenol undergoes multiple substitution with bromine. Benzene is monosubstituted only.

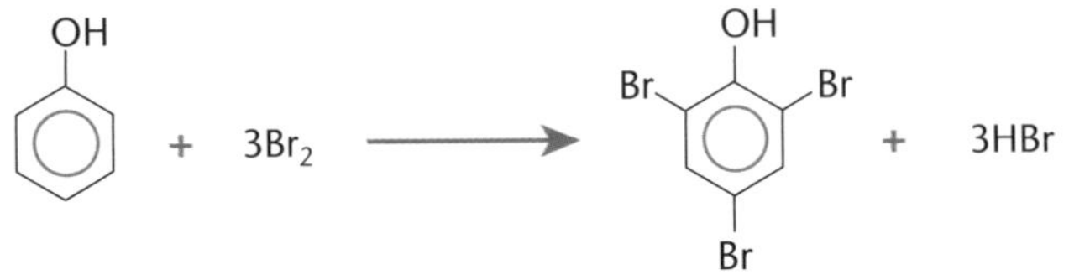

- The organic product 2,4,6-tribromophenol separates as a white solid.
- The bromine is decolourised.

Reaction with iron(III) chloride, $FeCl_3$

SALTERS M4

Phenols form a violet coloration with $FeCl_3$ solution.

- This reaction is used as a test for a phenol.
- This allows phenols to be distinguished from alcohols.

Formation of esters

SALTERS M4

Esters can be formed from an **alcohol** with a **carboxylic acid** in the presence of an acid catalyst (see page 100).

In phenols, the aryl ring reduces the electron density on the –OH group, reducing its effectiveness as a nucleophile. This means that phenols do **not** form esters in the same way as alcohols.

To form an ester from a phenol:

- an acyl chloride is used rather than the less reactive carboxylic acid (see page 105)
- alkaline conditions are used.

$$CH_3COCl + C_6H_5OH + NaOH \longrightarrow CH_3COOC_6H_5 + NaCl + H_2O$$

Progress check

1 Phenols are easily nitrated with nitric acid in the cold. With concentrated HNO_3, multiple substitution takes place.
Suggest an equation for the reaction that takes place.

1
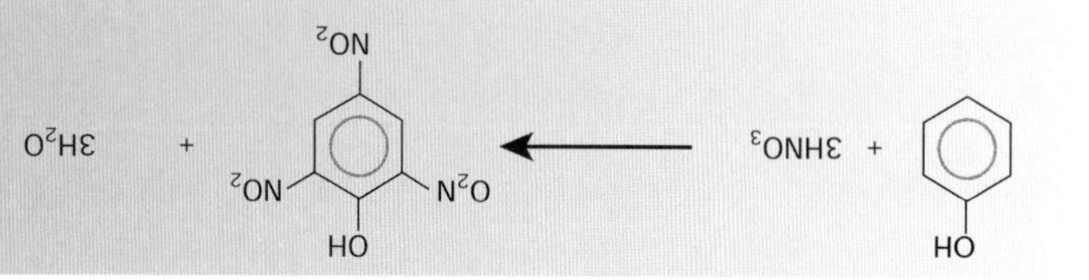

4.9 Amines

After studying this section you should be able to:

- *explain the relative basicities of ethylamine and phenylamine*
- *describe the reactions of primary amines with acids to form salts*
- *describe the formation of phenylamine by reduction of nitrobenzene*
- *describe the synthesis of an azo dye from phenylamine*
- *understand the origin of colour in azo dyes*
- *describe the reactions of amines with halogenoalkanes and acyl chlorides*
- *describe the preparation and reaction of amides*

LEARNING SUMMARY

Aliphatic and aromatic amines

OCR M4
SALTERS M4

Amines are organic compounds containing nitrogen derived from ammonia, NH_3. Amines are classified as primary, secondary or tertiary amines depending on how many of the hydrogen atoms in ammonia have been replaced by organic groups. The diagram below shows examples of aliphatic and aromatic amines.

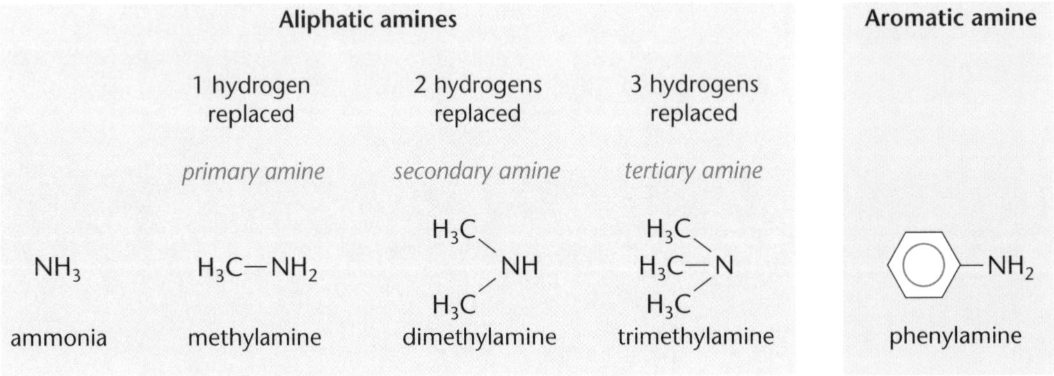

Amines in nature

Amines are found commonly in nature. They are weak bases and the shortest chain amines, e.g. ethylamine $C_2H_5NH_2$, smell of fish. Some diamines, such as putrescine $H_2N(CH_2)_4NH_2$, and cadaverine $H_2N(CH_2)_5NH_2$, are found in decaying flesh.

Polarity

Notice the similarity with ammonia.

The presence of an electronegative nitrogen atom in amines results in polar molecules. The diagram shows the polarity of the primary amine methylamine.

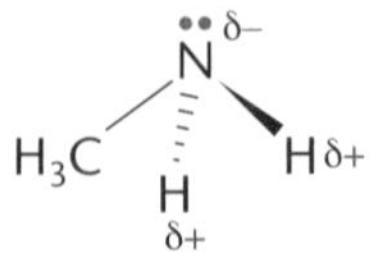

Physical properties of amines

As with other polar functional groups, the solubility of amines in water decreases with increasing carbon chain length as the non-polar contribution to the molecule becomes more important.

The amino group dominates the physical properties of short-chain amines.

Hydrogen bonding takes place between amine molecules, resulting in:

- higher melting and boiling points than alkanes of comparable relative molecular mass
- solubility in water – amines with fewer than six carbons mix with water in all proportions.

Amines as bases

OCR M4
SALTERS M4

Amines are weak bases because they only partially associate with protons:

$$RNH_2 + H^+ \rightleftharpoons RNH_3^+$$

The basic strength of an amine measures its ability to **accept** a proton, H^+. This depends upon:

- the size of the δ– charge on the amino nitrogen atom
- the availability of the nitrogen lone pair.

Amines are the organic bases.

The basicity of aliphatic and aromatic amines

Aliphatic amines are stronger bases than ammonia; aromatic amines are substantially weaker.

Amines can also act as ligands with transition metal ions forming complex ions.

KEY POINT

- Electron-donating groups, e.g. alkyl groups, **increase** the basic strength.
- Electron-withdrawing groups, e.g. C_6H_5, **decrease** the basic strength.

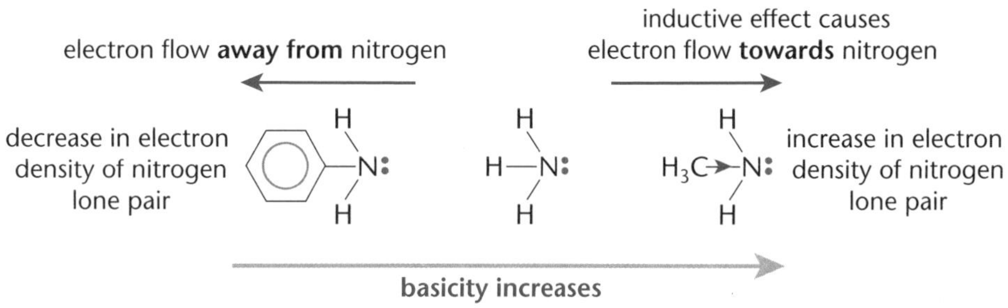

Organic ammonium salts

Amines react by the usual 'base reactions' producing organic ammonium salts.

A proton, H^+, is added to the amino nitrogen atom.

Amines are **neutralised by acids** forming salts.

In the example below, methylamine is neutralised by hydrochloric acid forming a primary **ammonium salt**.

$$CH_3NH_2 + HCl \longrightarrow CH_3NH_3^+Cl^-$$

methylammonium chloride

On evaporation of water, the primary ammonium salt crystallises out as an ionic compound.

Preparation of amines

OCR M4

Preparation of aromatic amines from nitroarenes

Other reducing agents can also be used:

- H_2/Ni catalyst
- $LiAlH_4$ in dry ether

$LiAlH_4$ is an almost universal reducing agent and works in most examples of organic reduction.

Aromatic amines can be prepared by **reducing** a nitroarene. For example, nitrobenzene is reduced by Sn in concentrated HCl to form phenylamine.

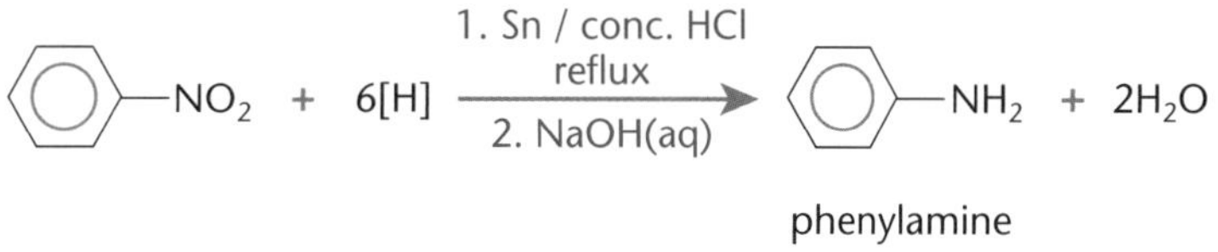

- Using tin and hydrochloric acid, the salt $C_6H_5NH_3^+Cl^-$ is formed.
- The amine is obtained from this salt by adding aqueous alkali.

The preparation of dyes from aromatic amines

OCR M4
SALTERS M5

Aromatic amines, such as phenylamine, are important industrially for the production of dyes. Modern dyes are formed in a two-stage synthesis:

- the aromatic amine is converted into a **diazonium salt**
- the diazonium salt is **coupled** with an aromatic compound such as phenol, forming an **azo dye**.

Formation of diazonium salts

An aromatic amine, such as phenylamine, forms a diazonium salt in the presence of nitrous acid, HNO_2, and hydrochloric acid.

- HNO_2 is unstable and is prepared *in situ* from $NaNO_2$ and HCl(aq).

$$NaNO_2 + HCl \longrightarrow HNO_2 + NaCl$$

- This mixture is then reacted with phenylamine. It is important to keep the temperature **below 10°C** because diazonium salts decompose above this temperature.

benzenediazonium chloride
(diazonium salt)

$$C_6H_5NH_2 + HNO_2 + HCl \longrightarrow C_6H_5N_2^+Cl^- + 2H_2O$$

Formation of azo dyes by coupling

The diazonium salt is coupled with a suitable aromatic compound **below 10°C** in **aqueous alkali** to produce an azo dye.

E.g. coupling of benzenediazonium chloride with phenol:

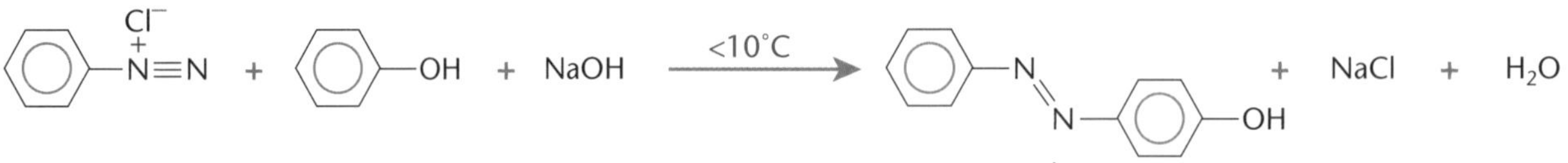

azo dye

- The azo dye is coloured.
- Coupling of a diazonium salt with different aromatic compounds forms dyes with different colours.

Using benzene for the synthesis of dyes

Benzene is used as the raw material for the synthesis of dyes.
A four-stage synthesis is shown below:

- benzene is nitrated to nitrobenzene (see page 111)
- nitrobenzene is reduced to phenylamine (see page 119)
- phenylamine is converted into a **diazonium salt**
- the diazonium salt is **coupled** with an aromatic compound such as phenol, forming an **azo dye**.

Colour in azo dyes

SALTERS M5

When energy is absorbed, the electron moves between electronic energy levels.

$\Delta E = hf$

Where ΔE = energy gap,

h = the Planck constant

f = frequency

(See page 73 for more details)

On pages 73–74, the origin of colour in transition elements is discussed. For organic compounds, the principles are the same.

When visible light strikes an organic compound that is coloured:

- an electron is promoted from a lower to a higher energy level
- **energy** of certain wavelengths is **absorbed**
- part of the visible spectrum has been removed and the **remaining wavelengths** are **reflected as colour**.

> **KEY POINT**
> A **chromophore** is part of an organic molecule responsible for its colour.

The ultraviolet region of the spectrum has more energy than the visible region – the gaps between its energy levels are greater.

- Common chromophores include C=C, C=O, N=N, an aromatic ring, and features with lone pairs such as OH, NH_2.
- Colour is produced when the energy gap between energy levels matches the energy from part of the visible spectrum.
- For many organic molecules containing chromophores, this energy gap is large and matches energies in the ultraviolet region of the spectrum.
- Delocalisation reduces this energy gap. When the delocalisation is such that the energy gap matches energies in the visible part of the spectrum, the organic compound is coloured.

The colour of an organic compound results from the **transfer** of an **electron** between energy levels in a delocalised system.

Colourless

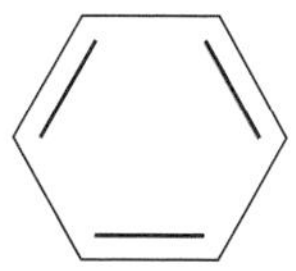

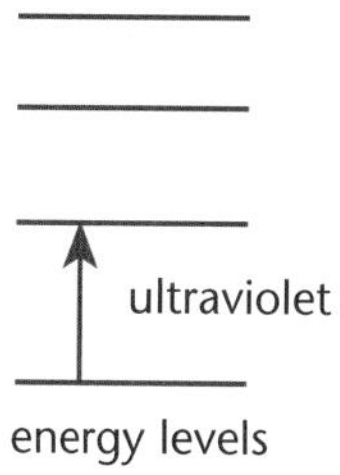

Kekulé structure of benzene
Some delocalisation but electron energy gap is large and matches ultraviolet region of spectrum

Coloured

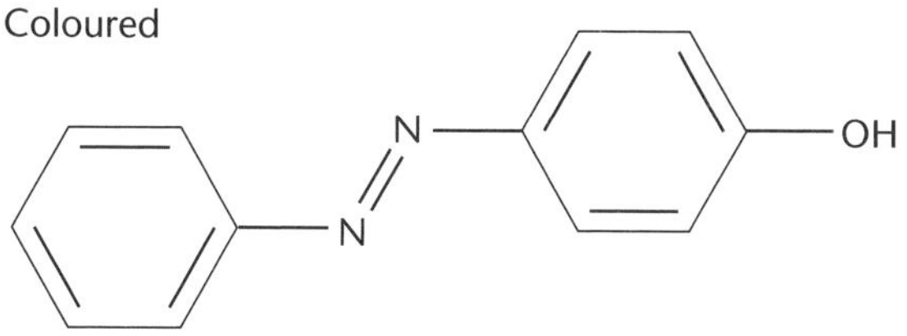

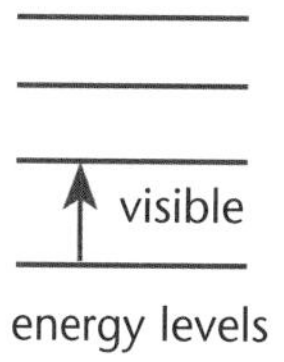

Azo dye has more delocalisation
(note the alternating double and single bonds)
Energy gap is smaller and matches visible part of spectrum

The key to colour in organic compounds is a delocalised arrangement of alternating double and single bonds, such as that in an azo dye.

Attaching dyes to fibres

Dyes attach themselves to fibres by:

- intermolecular bonds
- ionic attractions
- covalent bonding.

Modifying dyes

Chemists modify dyes by introducing different functional groups. These may:

- modify the chromophore
- affect the solubility of the dye
- allow the dye to bond to fibres.

Reactions of amines with halogenoalkanes

OCR M4

Key points from AS

- **Nucleophilic substitution reactions of halogenoalkanes**
  Revise AS pages 116–118

Formation of a primary amine by nucleophilic substitution

Amines act as **nucleophiles** with organic compounds such as halogenoalkanes.

Halogenoalkanes react with **excess ammonia** in hot ethanol, forming a primary amine. The reaction with bromomethane is shown below:

$$CH_3Br + 2NH_3 \longrightarrow CH_3NH_2 + NH_4^+Br^-$$

The reaction takes place in two steps:

- Nucleophilic substitution with ammonia forms an alkylammonium ion:

$$CH_3Br + NH_3 \longrightarrow CH_3NH_3^+ + Br^-$$

- Proton transfer with ammonia forms a primary amine:

$$CH_3NH_3^+Br^- + NH_3 \rightleftharpoons CH_3NH_2 + NH_4^+Br^-$$

Amides

SALTERS M4

Amides

Primary amides have the general formula $RCONH_2$ – R is an alkyl group.

The first six members of the **amides** homologous series are shown below.

structural formula	*molecular formula*	*name*
$HCONH_2$	CH_3NO	methanamide
CH_3CONH_2	C_2H_5NO	ethanamide
$CH_3CH_2CONH_2$	C_3H_7NO	propanamide
$CH_3CH_2CH_2CONH_2$	C_4H_9NO	butanamide
$CH_3CH_2CH_2CH_2CONH_2$	$C_5H_{11}NO$	pentanamide
$CH_3CH_2CH_2CH_2CH_2CONH_2$	$C_6H_{13}NO$	hexanamide

Types of amide

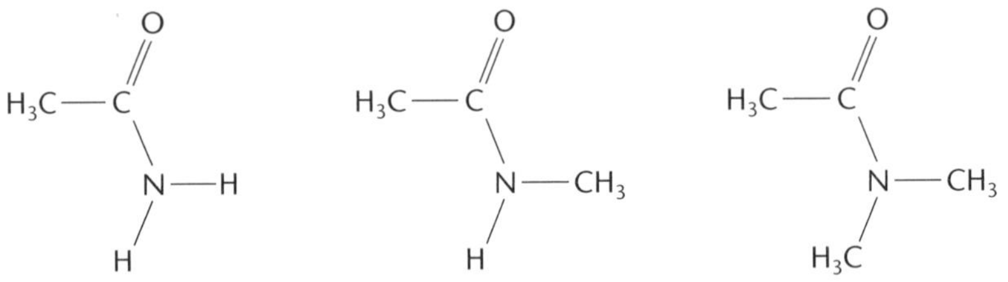

primary amide secondary amide tertiary amide

Preparation of amides

The amide group is stable and relatively unreactive. Preparation directly from a carboxylic acid is difficult (although this type of reaction is essentially what happens when peptides form from amino acids – see pages 128 and 135).

Amides are generally prepared by the reaction of ammonia and amines with **acyl chlorides** or with acid anhydrides – these are much more reactive than carboxylic acids (see pages 104–105).

- Reaction with ammonia produces a **primary amide**.

  $CH_3COCl + NH_3 \longrightarrow + \mathbf{CH_3CONH_2} + HCl$

  (Then $NH_3 + HCl \longrightarrow NH_4Cl$)

- Reaction with a primary amine produces a **secondary amide**.

  $CH_3COCl + CH_3NH_2 \longrightarrow + \mathbf{CH_3CONHCH_3} + HCl$

  (Then $NH_3 + HCl \longrightarrow NH_4Cl$)

The acyl chloride is often prepared initially from a carboxylic acid. The acyl chloride can then be reacted with ammonia or amines to form amides.

An amide can be prepared from a carboxylic acid by first converting the carboxylic acid into an acyl chloride (see pages 104–105 for more details).

Hydrolysis of amides

Amides are stable organic compounds. They can be hydrolysed but they require much stronger conditions than esters. Alkaline hydrolysis requires prolonged refluxing with 6 mol dm^{-3} NaOH(aq).

Acid hydrolysis

Acid hydrolysis of amides form a carboxylic acid and an ammonium salt:

$CH_3CONH_2 + H_2O + HCl \longrightarrow CH_3COOH + NH_4^+Cl^-$

Alkaline hydrolysis

Alkaline hydrolysis of amides form an acid salt and ammonia:

$CH_3CONH_2 + NaOH \longrightarrow CH_3COO^-Na^+ + NH_3$

Progress check

1 Explain why ethylamine is a stronger base than phenylamine.

2 (a) Write equations for the conversion of:
(i) benzene to nitrobenzene
(ii) nitrobenzene to phenylamine
(iii) phenylamine to benzenediazonium chloride.

3 (a) Write down the formula of the organic product formed between:
(i) bromoethane and excess ammonia
(ii) ammonia and excess bromoethane.

1 The ethyl group has a positive inductive effect which reinforces the electron density of the nitrogen atom of the amine. This lone pair on the nitrogen will attract protons more strongly.
The phenyl group has a negative inductive effect which reduces the electron density on the nitrogen atom of the amine. This lone pair on the nitrogen will attract protons less strongly.

2 (a) (i) $C_6H_6 + HNO_3 \longrightarrow C_6H_5NO_2 + H_2O$ (ii) $C_6H_5NO_2 + 6[H] \longrightarrow C_6H_5NH_2 + 2H_2O$
(iii) $C_6H_5NH_2 + HNO_2 + HCl \longrightarrow C_6H_5N_2^+Cl^- + 2H_2O$

3 (a) (i) $C_2H_5NH_2$ (ii) $(C_2H_5)_4N^+Br^-$

Sample question and model answer

This question looks at the structure and reactions of benzene.

(a) The average enthalpy of hydrogenation of a single C–C bond is –120 kJ mol^{-1}.

(i) Assuming that benzene consists of a ring with three separate double bonds, predict the enthalpy change for the hydrogenation of benzene to form cyclohexane.

3×-120 kJ mol^{-1} = -360 kJ mol^{-1} ✓

> This is an easy mark.
> Don't forget to include the sign!

(ii) The actual enthalpy of hydrogenation of benzene is –205 kJ mol^{-1}. Using this information and your answer to (i), what conclusion can be drawn about the stability of the benzene ring? Use an enthalpy level diagram to illustrate your answer.

Benzene has extra stability arising from delocalisation of -electrons ✓

> An enthalpy diagram is a good way of showing the delocalisation energy of delocalised benzene.
> 1 mark here is awarded for the relative positions of the delocalised and Kekulé structures of benzene.

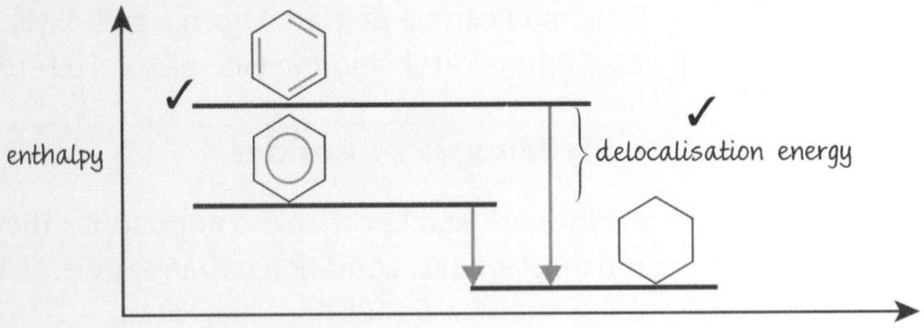

[3]

(b) Benzene can be nitrated to form nitrobenzene. Give the reagents, the equation and the conditions for this reaction.

Reagent(s): concentrated HNO_3/H_2SO_4 ✓

Equation: $C_6H_6 + HNO_3 \longrightarrow C_6H_5NO_2 + H_2O$ ✓

Conditions: warm to 55°C ✓ [3]

> This is standard bookwork that must be learnt.
> Other reactions could have been chosen (e.g. bromination).
> Remember that 'reagents' are the chemicals 'out of the bottle' that are reacted with the organic compound.
> You must give the full name or formula for any 'reagent'.

(c) For the nitration of benzene:

(i) write the formula of the electrophile

NO_2^+ ✓

(ii) write the equation to show the formation of the electrophile

$HNO_3 + H_2SO_4 \longrightarrow NO_2^+ + HSO_4^- + H_2O$ ✓

> More standard bookwork.
> An alternative response here would be:
> $HNO_3 + 2H_2SO_4 \rightarrow NO_2^+ + 2HSO_4^- + 2H_2O$

(iii) state the type of mechanism

electrophilic substitution ✓

(iv) outline the mechanism.

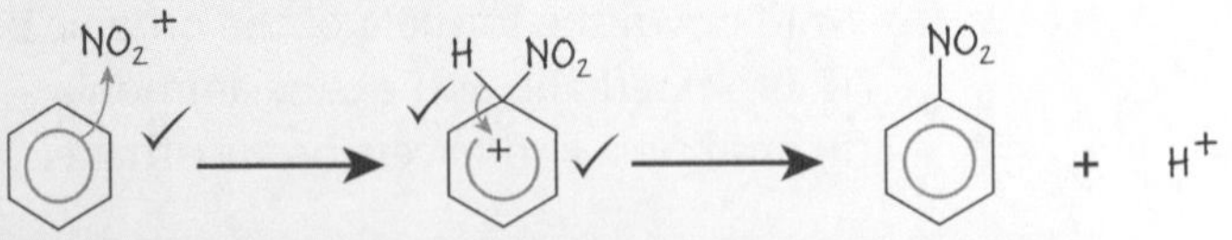

The sulfuric acid catalyst is then regenerated:

$H^+ + HSO_4^- \longrightarrow H_2SO_4$ ✓

[7]

[Total:13]

> Notice how precise you need to be to score all three marks:
> - arrow to electrophile ✓
> - correct intermediate ✓
> - arrow on C–H for loss of H^+ ✓.

Practice examination questions

1 Compounds **A**, **B** and **C** are structural isomers, each containing a carbonyl group.

The compounds have a relative molecular mass of 72 and the following percentage composition by mass: C, 66.7%; H, 11.1%; O, 22.2%.

(a) Calculate the molecular formula of the isomers. [2]

(b) What are the structural formulae of **A**, **B** and **C**? [3]

(c) Name a reagent that reacts with **A**, **B** and **C**. State the observation. [2]

(d) Name a reagent that reacts with only two of the isomers **A**, **B** and **C**. Identify which isomers react and state the observation. [3]

(e) One of the isomers **A**, **B** and **C** is reacted with $NaBH_4$ to form a product that has optical isomers. Identify this product and the isomer that has been reacted with $NaBH_4$. [2]

[Total: 12]

2 (a) Compound **D**, $CH_3CH_2COOCH_2CH_3$, is an ester.

(i) Name compound **D**.

(ii) Compound **D** was hydrolysed by heating with aqueous sodium hydroxide. An alcohol **E** was formed, together with another organic product, **F**. Give the structural formulae of compounds **E** and **F**. [4]

(b) Compound **G** is a di-ester formed by the reaction of ethane-1,2-diol and methanoic acid.

(i) Draw the structural formula of compound **G**.

(ii) Identify a substance that could catalyse this reaction. [2]

(c) Compound **H** is a tri-ester. Complete and balance the equation below for the formation of the tri-ester **H** by esterification.

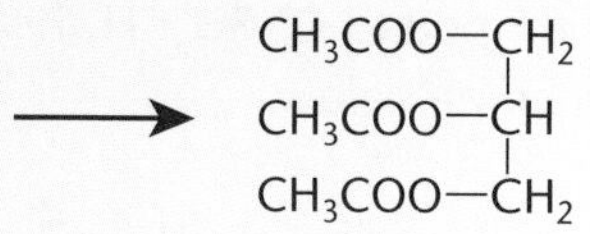

tri-ester H

[3]

[Total: 9]

3 An aromatic hydrocarbon has a relative molecular mass of 106 and has the following composition by mass: C, 90.56%; H, 9.44%.

(a) (i) Deduce the molecular formula of the aromatic hydrocarbon.

(ii) Draw structures for all the structural isomers of this aromatic hydrocarbon. [7]

(b) In the presence of a sulfuric acid, one of these isomers, **I**, reacts with nitric acid to give only one mononitro product **J**.

(i) Deduce which of the isomers in (a)(ii) is isomer **I**.

(ii) Draw the structure of **J**. [2]

[Total: 9]

Practice examination questions (continued)

4 Methylbenzene can be converted into an azo dye using the reaction scheme below.

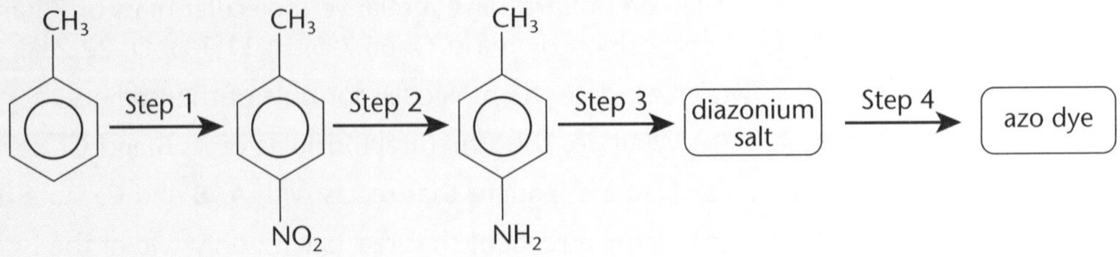

(a) For Step 1:

(i) name the mechanism

(ii) state the reagents required. [2]

(b) For the reduction in Step 2:

(i) state the reagents required

(ii) write a balanced equation (use [H] to represent the reducing agent). [3]

(c) For Step 3:

(i) state the reagents required and the essential conditions

(ii) show the structure of the diazonium salt formed (the functional group should be displayed with charges clearly shown). [3]

(d) For Step 4:

(i) give the name and formula of a suitable organic compound for coupling with the diazonium salt

(ii) show the structure of the azo dye formed. [2]

[Total: 10]

5 Compound **C**, $CH_3CH_2CH_2COOCH(CH_3)_2$, can be prepared using the reaction scheme below.

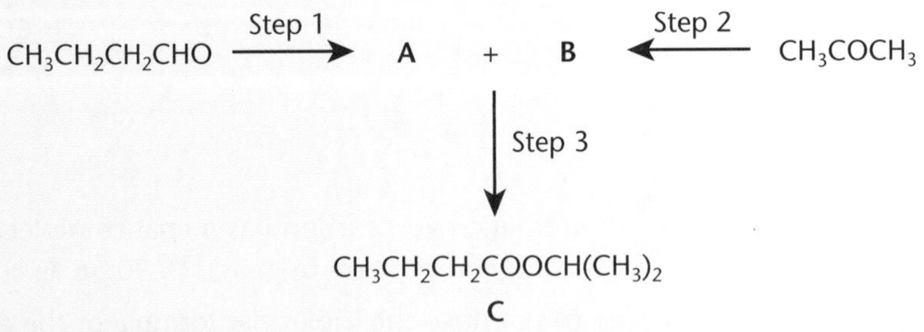

(a) Show the structural formulae of compounds **A** and **B**. [2]

(b) For each of the three steps, give the reagents and conditions and name the type of reaction. [9]

[Total: 11]

Chapter 5

Polymers, analysis and synthesis

The following topics are covered in this chapter:

- *Amino acids*
- *Polymers, proteins and nucleic acids*
- *Chromatography*
- *Analysis*
- *NMR spectroscopy*
- *Organic synthetic routes*

5.1 Amino acids

After studying this section you should be able to:

- *describe the acid–base properties of amino acids and the formation of zwitterions*
- *explain the formation of polypeptides and proteins as condensation polymers of amino acids*
- *describe the acid hydrolysis of proteins and peptides*

LEARNING SUMMARY

Amino acids

OCR M4
SALTERS M4

A typical amino acid is an organic molecule with both acidic and basic properties comprising:

- a basic amino group, $-NH_2$
- an acidic carboxyl group, –COOH.

There are 22 naturally occurring amino acids.

Each amino acid has a unique organic R group or side chain. The formula of a typical amino acid is shown below.

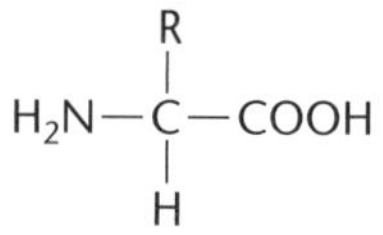

Examples of some amino acids are shown below:

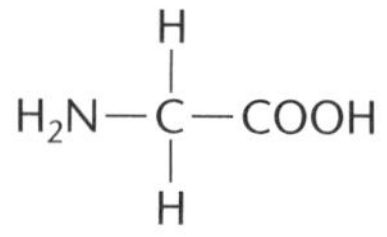

R = H
glycine

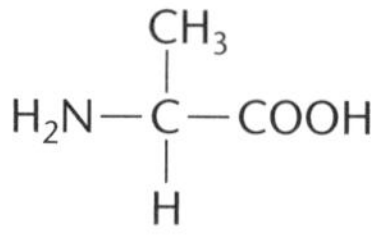

$R = CH_3$
alanine

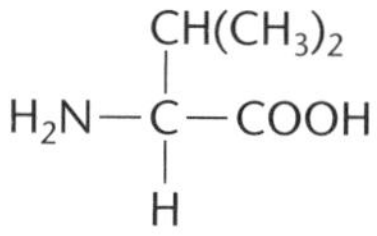

$R = CH(CH_3)_2$
valine

Acid–base properties of amino acids

OCR M4
SALTERS M4

Isoelectric points and zwitterions

Each amino acid has a particular pH called the **isoelectric point** at which the overall charge on an amino acid molecule is zero.

Examples of isoelectric points

amino acid	aspartic acid	glycine	histidine	arginine
isoelectric point	3.0	6.1	7.6	10.8

At the isoelectric point, the amino acid exists in equilibrium with its zwitterion form.

At the isoelectric point, an amino acid exists as a zwitterion:

- the **carboxyl** group **has donated** a proton to the **amino** group, which form a positive NH_3^+ ion.

A **zwitterion** is a dipolar ion with both positive and negative charges in different parts of the molecule.

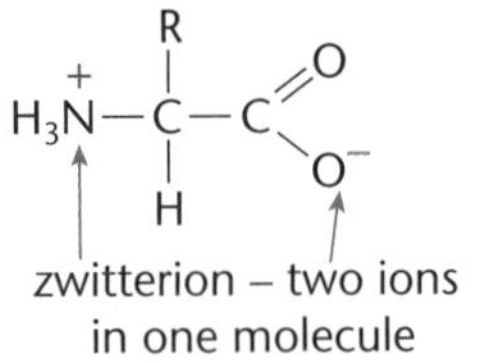

Amino acids as bases

In strongly **acidic** conditions a **positive ion** forms:

- an amino acid behaves as a **base**
- the COO^- ion gains a proton.

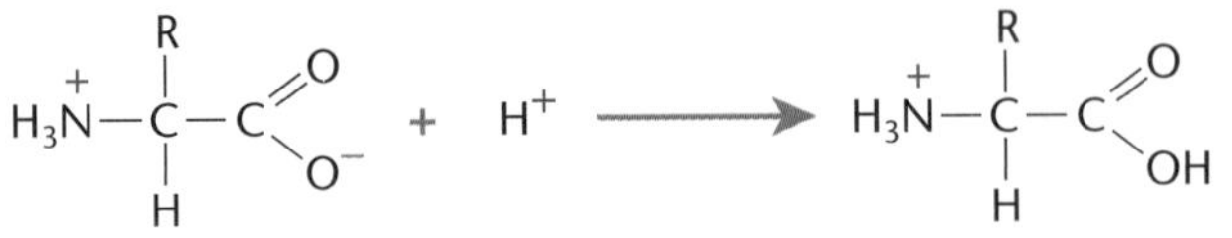

positive ion

Because of their reactions with strong acids and strong bases, amino acids act as buffers and help to stabilise the pH of living systems.

Amino acids as acids

In strongly **alkaline** conditions a **negative ion** forms:

- an amino acid behaves as an **acid**
- the NH_3^+ ion loses a proton.

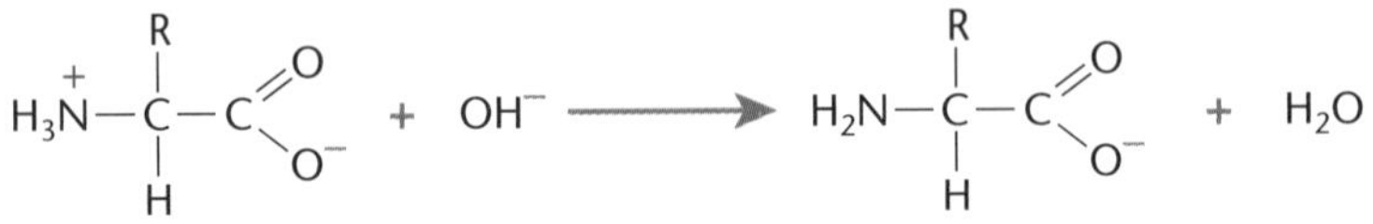

negative ion

> **KEY POINT**
> - At the **isoelectric point**, the amino acid is **neutral**.
> - At a pH more **acidic** than the isoelectric point, the amino acid forms a **positive ion**.
> - At a pH more **alkaline** than the isoelectric point, the amino acid forms a **negative ion**.

Physical properties of amino acids

OCR M4
SALTERS M4

Solid amino acids

Solid amino acids have higher melting points than expected from their molecular masses and structure. This suggests that amino acids crystallise in a giant lattice with strong electrostatic forces between the **zwitterions**.

Optical isomers

With the exception of aminoethanoic acid (*glycine*), H_2NCH_2COOH, all amino acids have a chiral centre and are optically active.

Polypeptides and proteins

OCR M4
SALTERS M4

In nature, individual amino acids are linked together in chains as **polypeptides** and **proteins** (see page 135).

The diagram below shows the condensation of the amino acids glycine and alanine to form a **dipeptide**. The amino acids are bonded together by a **peptide link**.

A protein is formed by condensation polymerisation of amino acids. See page 135.

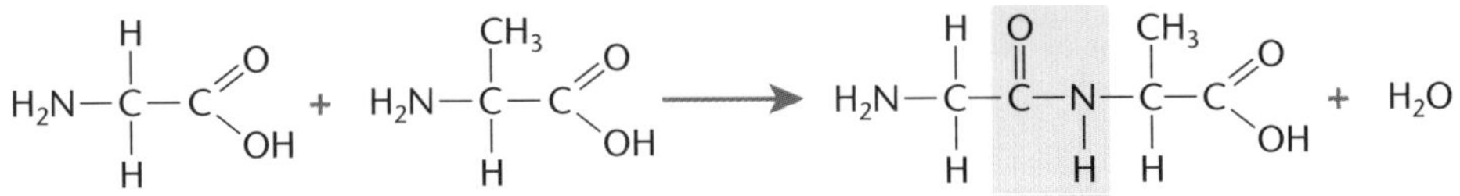

glycine *alanine* *peptide link*

A polypeptide is the name given to a short chain of amino acids linked by peptide bonds.

A protein is simply the name given to a long-chain polypeptide.

Each peptide link forms:

- between the **carboxyl group** of glycine and the **amino group** of alanine
- with loss of a water molecule in a **condensation reaction**.

Further condensation reactions between amino acids build up a **polypeptide** or **protein**.

- For each amino acid added to a protein chain, one water molecule is lost.
- Most common proteins contain more than 100 amino acids.
- Each protein has a unique sequence of amino acids and a complex three-dimensional shape, held together by intermolecular bonds, including hydrogen bonds.

Hydrolysis of proteins

OCR M4
SALTERS M4

Hydrolysis breaks down a protein into its separate amino acids.

- The protein is refluxed with 6 mol dm^{-3} HCl(aq) for 24 hours.
- The resulting solution is neutralised.

The equation below shows the hydrolysis of a dipeptide.

Proteins can also be hydrolysed with hot aqueous alkali.

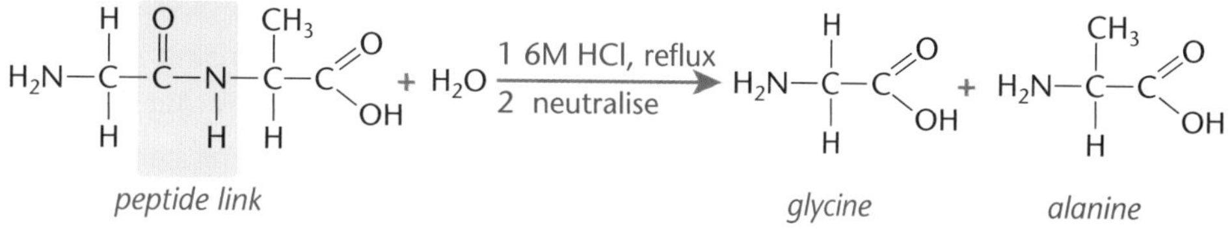

- Hydrolysis of a protein is the reverse process to the condensation polymerisation that forms proteins.
- Biological systems use enzymes to catalyse this hydrolysis, which takes place at body temperature without the need for acid or alkali.

Identifying the amino acids in a protein

To determine the amino acids present in a protein, the protein is first boiled with 6 mol dm^{-3} hydrochloric acid. The amino acids formed are separated using paper chromatography and made visible by spraying the paper with ninhydrin.

Each amino acid moves a different distance on the chromatography paper, making for easy identification of the amino acids in the protein.

Progress check

1 The isoelectric point of serine (R = $-CH_2OH$) is 5.7. Draw the form of the molecule in aqueous solutions of pH 3.0, pH 5.7 and pH 10.0.

2 Draw the structure of the tripeptide with the sequence alanine–serine–aspartic acid (alanine: R = $-CH_3$; aspartic acid: R = $-CH_2COOH$).

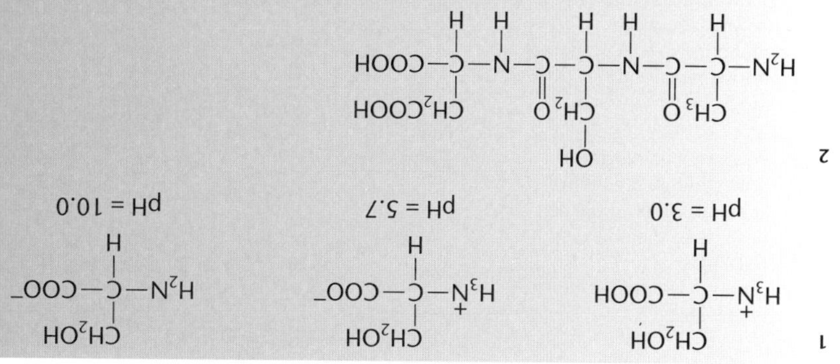

5.2 Polymers, proteins and nucleic acids

After studying this section you should be able to:

LEARNING SUMMARY

- *describe the characteristics of addition polymerisation*
- *describe the characteristics of condensation polymerisation in polypeptides, proteins, polyamides and polyesters*
- *discuss the disposal of polymers*
- *describe primary, secondary and tertiary structures of protein*
- *describe the characteristics of enzyme catalysis*
- *describe the structure and functions of DNA and its importance in directing protein synthesis*

During the study of alkenes in AS Chemistry, you learnt about addition polymers. For A2 Chemistry, addition polymerisation is reviewed and compared with condensation polymerisation.

Monomers and polymers

OCR M4
SALTERS M4

A **polymer** is a compound comprising very large molecules that are multiples of simpler chemical units called **monomers**.

Monomers are small molecules that can combine together to form a single large molecule, called a **polymer**.

KEY POINT

Two processes lead to formation of a polymer.

- **Addition polymerisation** – monomers react together forming the polymer **only**. There are no by-products.
- **Condensation polymerisation** – monomers react together forming the polymer **and** a simple compound, usually water.

Addition polymerisation

OCR M4
SALTERS M4

Key points from AS

- **Addition polymerisation of alkenes**
  Revise AS pages 110–111

Make sure that you can show a repeat unit of a polymer from the monomer (and a monomer from a repeat unit).

Poly(propenamide) and poly(ethenol) are highly water-absorbent polymers because of the ability of the polar –OH and –$CONH_2$ groups to hydrogen bond with water.

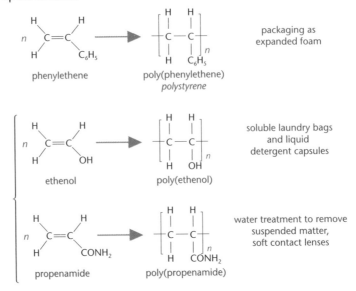

In addition polymerisation:

- the monomer is an **unsaturated** molecule with a double **C=C** bond
- the double bond is **lost** as the **saturated** polymer forms.

Many different addition polymers can be formed using different monomer units based upon alkenes.

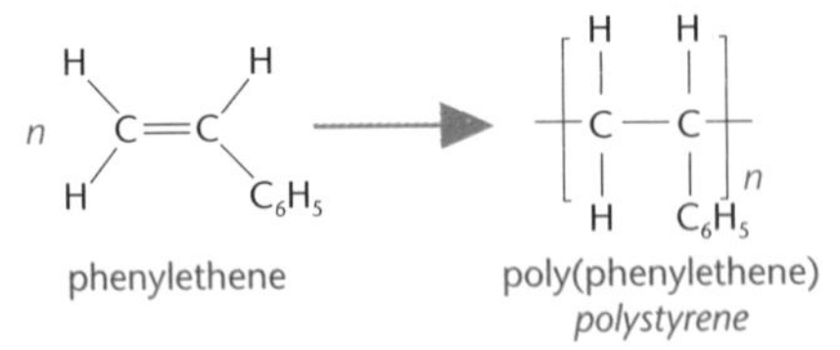

packaging as expanded foam

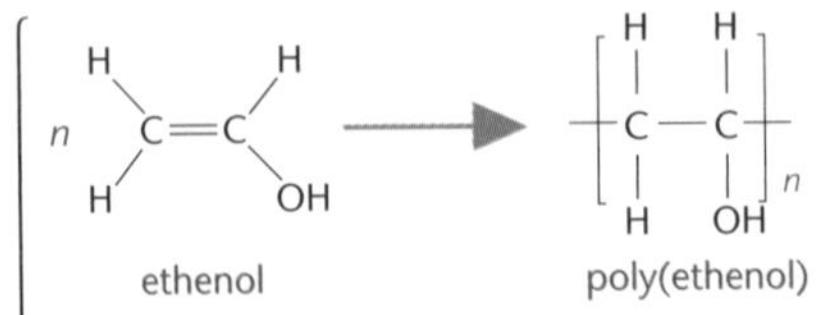

soluble laundry bags and liquid detergent capsules

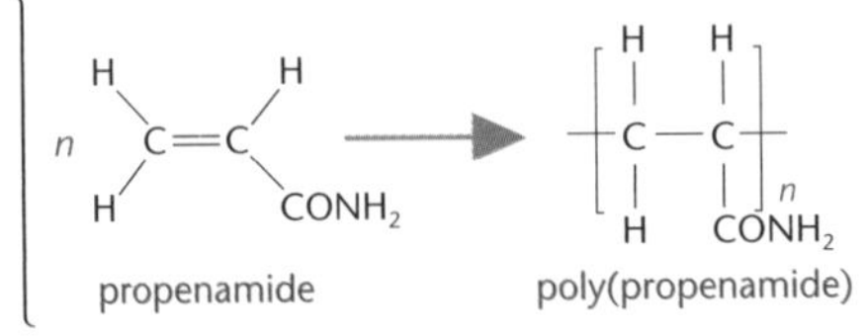

water treatment to remove suspended matter, soft contact lenses

Properties of monomers and addition polymers

The monomers are volatile liquids or gases. Polymers are solids.
This difference can be explained in terms of van der Waals' forces.

- The van der Waals' forces acting between the large polymer molecules are much stronger than those acting between the much smaller monomer molecules.

Addition polymers of hydrocarbon monomers and of halogenated monomers are very stable and inert, e.g. poly(ethene), poly(propene) and poly(phenylethene) (polystyrene), poly(chloroethene) (PVC) and poly(tetrafluoroethene) (PTFE). These polymers are insoluble in solvents and are non-biodegradable.

SALTERS M4

Branching, crystallinity and properties

The flexibility and softening of polymers such as LDPE and HDPE depend upon temperature, branching and crystallinity.

Crystallinity

HDPE is more crystalline than LDPE.
The polymer chains in HDPE can align with one another more easily.

Crystallinity is the regular packing of the chains, due to the regular structure of the polymer. The lower the temperature, the more effective the intermolecular bonds.

- A polymer softens and loses its crystallinity above its 'crystalline melting temperature', T_m
- A polymer becomes brittle below its glass transition temperature, T_g.

Chain length and branching

HDPE has non-branched chains.
LDPE has branched chains.
HDPE has stronger intermolecular forces and has more strength than LDPE.

- As chain length increases, there are more intermolecular bonds, leading to greater strength.
- With branched chains, the polymer chains cannot get as close together as in non-branched chains and intermolecular bonds will be weaker.
- Flexibility depends on the ability of the polymer chains to slide over each other and this is more evident in polymers with branched chains.

LDPE is flexible.
HDPE is rigid.

Modifying properties

The properties of all materials depend on their structure and bonding.

Chemists can modify the properties of a polymer to meet particular needs by:

- cold-drawing, to make the structure more crystalline
- copolymerisation – producing mixed polymers by using more than one monomer
- use of plasticisers, which change T_g and increase flexibility.

Condensation polymerisation

OCR M4
SALTERS M4

In condensation polymerisation, the formation of a bond between monomer units also produces a small molecule such as H_2O or HCl.

Condensation polymers can be divided into natural polymers and man-made (synthetic) polymers.

- Natural polymers in living organisms include proteins, cellulose, rayon and DNA.
- Man-made polymers include synthetic fibres such as polyamides (e.g. *nylon*) and polyesters (e.g. *terylene*).

Polyamides

Nylon, proteins and polypeptides are all polyamides.

Proteins and polypeptides are natural condensation polymers. The link between the amino acid monomer units is usually described as a **peptide link** but chemically this is identical to an **amide** group. Hence polypeptides and proteins are **natural polyamides**.

Nylon-6,6 was the first man-made condensation polymer and was synthesised as an artificial alternative to natural protein fibres such as wool and silk.

Compare the use of two monomers in the production of synthetic nylon-6,6 with the natural condensation polymerisation using amino acids only.

The principle used was to mimic the natural polymerisation process above but, instead of using an amino acid monomer with two different functional groups, **two** chemically different monomers are usually used:

- a dicarboxylic acid **A**
- a diamine **B**

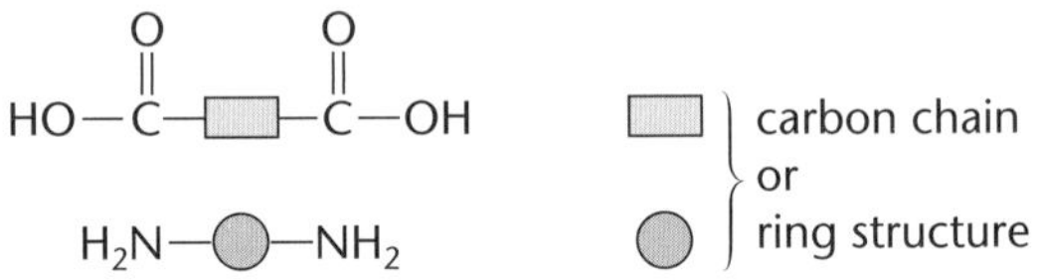

Many different polyamides can be made using different carbon chains or rings which bridge the double functional group.

Each diamine molecule bonds to a dicarboxylic acid molecule with loss of water molecule.

- the two monomers **A** and **B** join alternately: **–A–B–A–B–A–B–A–B–**

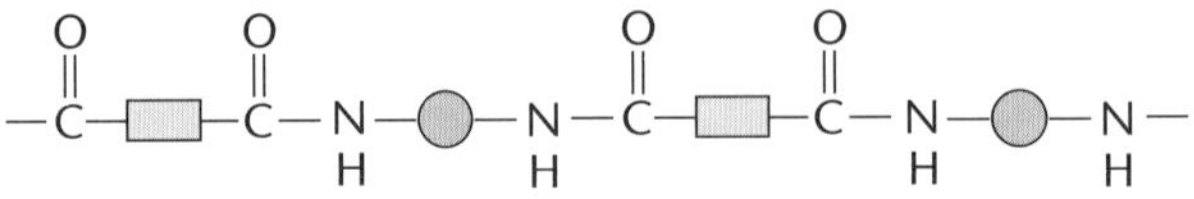

Polyamides can also be made using a diacyl chloride instead of a dicarboxylic acid.

- The greater reactivity of an acyl chloride results in easier polymerisation.
- Hydrogen chloride is lost instead of water.

Formation of nylon-6,6 from its monomers

Nylon-6,6 gets its name from the number of carbon atoms in each monomer (diamine first).

The diamine has 6 carbon atoms.

The dicarboxylic acid has 6 carbon atoms.

Hence: nylon-6,6.

The diacyl chloride for preparing nylon-6,6 would be $ClOC(CH_2)_4COCl$.

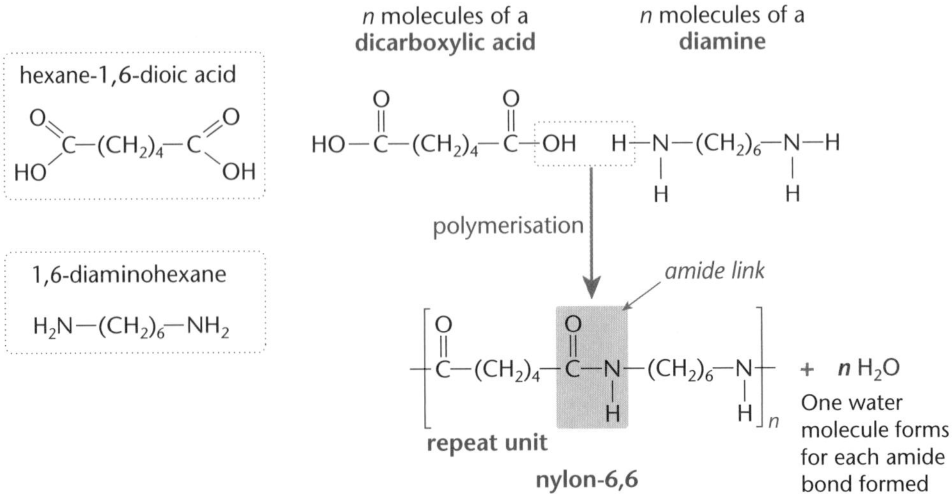

Formation of Kevlar from its monomers

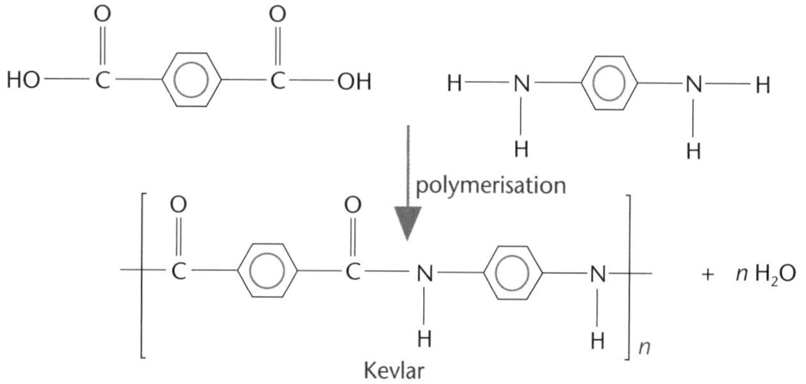

Kevlar

Other polyamides include Kevlar, one of the hardest materials known. Kevlar is used for bulletproof vests, belts for radial tyres, cables and reinforced panels in aircraft and boats.

Polyesters

Learn the principle behind condensation polymerisation.

In exams, you may be required to predict structures from unfamiliar monomers.

However, the principle is the same.

Polyesters are polymers made by a condensation reaction between monomers with formation of an ester group as the linkage between the molecules.

The first man-made polyester produced was *Terylene*. As with polyamides, polyesters are used as fibres for clothing.

As with artificial polyamides, man-made polyesters are usually made from **two** different monomers:

- a dicarboxylic acid
- a diol

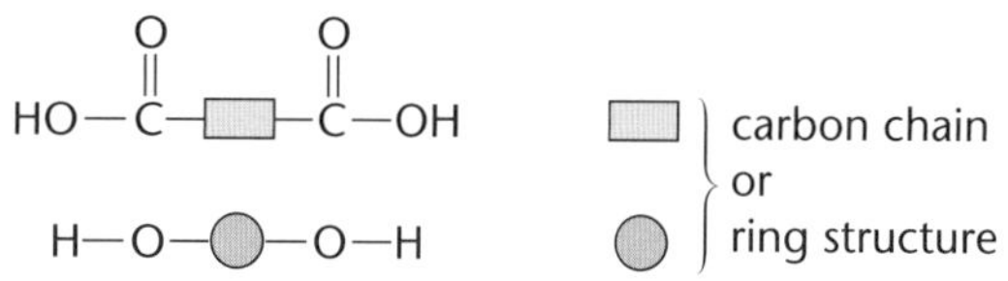

Each diol molecule bonds to a dicarboxylic acid molecule with loss of a water molecule.

Many different polyesters can be made by using different carbon chains or rings bridging the double functional group.

This is the same basic principle as for polyamides.

Formation of Terylene from its monomers

'Terylene' is a trade name for the polymer 'polyethylene terephthalate' (PET). PET is used to make the plastic for many drink bottles.

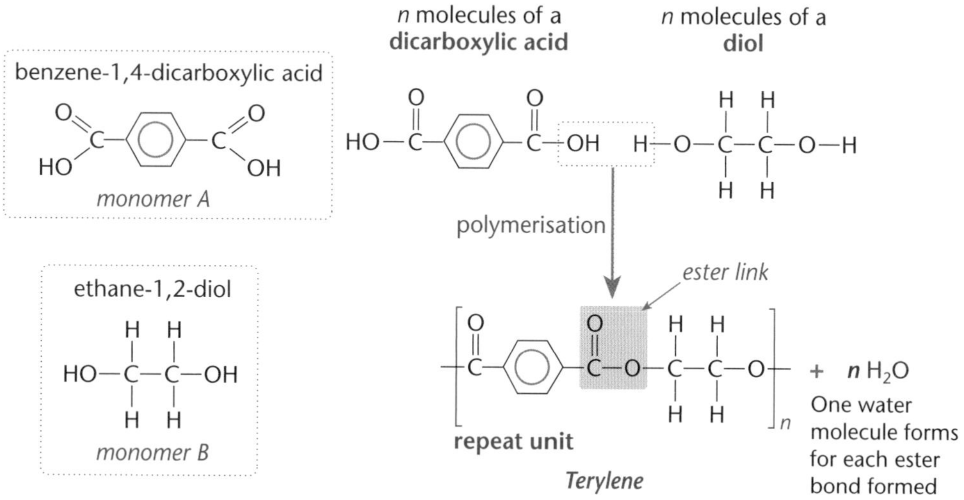

Disposal of polymers

OCR M4
SALTERS M4

Problems with addition polymers (polyalkenes)

Addition polymers are non-polar and chemically inert.
This creates potential environmental problems during disposal of polymers.
Disposal by landfill causes long-term problems.

- Addition polymers are **non-biodegradable** and take many years to break down.

Disposal by burning can produce toxic fumes.

- Depolymerisation produces poisonous monomers.
- Disposal of poly(chloroethene) (PVC) by incineration can lead to the formation of very toxic dioxins if the temperature is too low.

Solid domestic and industrial waste contains a high percentage of polymers.

Disposal requires waste management strategies such as:

- incineration – to reduce waste bulk and to generate energy
- recycling – to preserve natural oil-produced finite resources produced from oil.

Condensation polymers

Condensation polymers are polar.

- Condensation polymers are broken down naturally by acid and alkaline **hydrolysis** into their monomer units. They are therefore biodegradable and easier to dispose of than addition polymers.
- They may be photodegradable as the C=O bond absorbs radiation.

Chemists are minimising environmental waste by developing degradable polymers, similar in structure to poly(lactic acid).

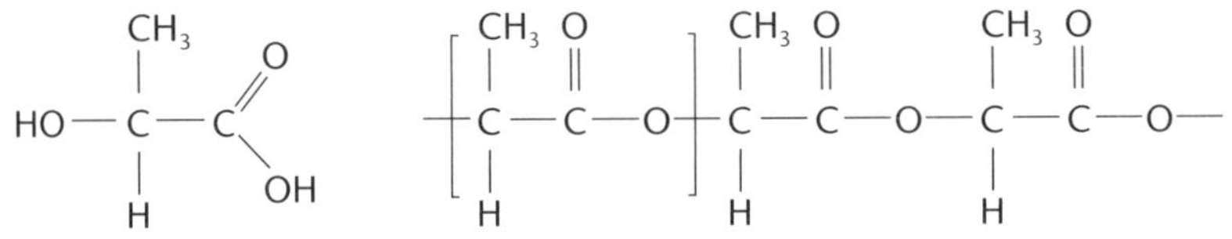

lactic acid **repeat unit** poly(lactic acid)

Poly(lactic acid) is used for waste sacks, packaging, disposable eating utensils and medical applications such as internal dissolvable stitches.

SALTERS M4

'Green polymers'

The principles of green chemistry are important in the manufacture, use, recycling and eventual disposal of polymers. They involve:

- minimising any hazardous waste during the production of raw materials and their resulting polymers to reduce any negative impact on the environment
- reducing carbon emissions resulting from the 'life cycle' of a polymer
- recycling to produce energy and chemical feedstocks.

Progress check

1 (a) Draw the structure of the monomer needed to make poly(tetrafluoroethene).
(b) Draw the repeat unit of the polymer perspex, made from the monomer methyl 2-methylpropenoate, shown below.

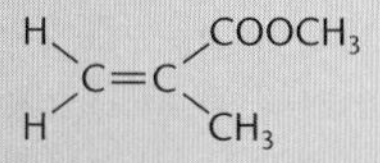

2 (a) Show the structures of the two monomers needed to make nylon-4,6.
(b) Draw a short section of nylon 4,6 and show its repeat unit.

3 Draw the structures of the two monomers needed to make the polyester shown below.

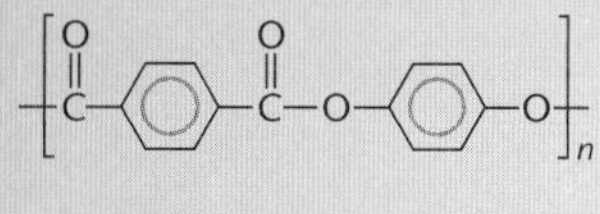

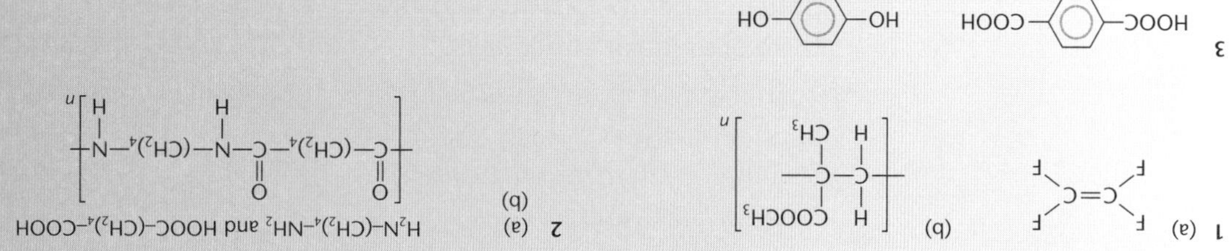

Proteins and polypeptides

OCR M4
SALTERS M4

Proteins and polypeptides

A protein is a chain of many amino acids linked together with 'peptide bonds' formed by **condensation polymerisation**.

Each amino acid molecule has both **amino** and **carboxyl** groups.
A **peptide** bond forms, with loss of a water molecule, between:

- the **amino** group of one amino acid molecule and
- the **carboxyl** group of another amino acid molecule.

The diagram below shows how amino acid molecules are linked together during condensation polymerisation.

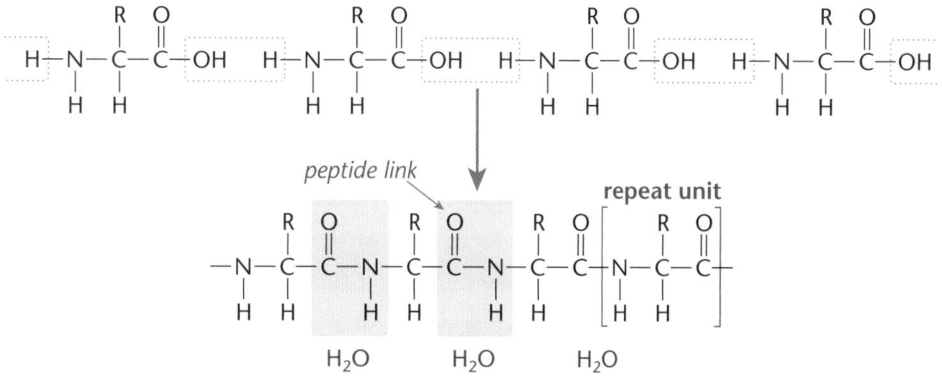

Although a 'repeat unit' is shown on the diagram, the R group may be different in each unit. There are over 20 different amino acids, each having a different R group.

- Most common proteins contain more than 100 amino acids.
- Each protein has a unique sequence of amino acids and a complex three-dimensional shape, held together by intermolecular bonds including hydrogen bonds.

SALTERS M4

Primary structure

Just think of how many 100 letter words you could make out of 20 letters of the alphabet!

The **primary** structure of a protein is the actual **sequence** of **amino acids** joined by **peptide** links.

e.g. ....ala–gly–lys–cys–asn–arg–gly–leu–try–val–............

- There are countless possible sequences, with over 20 different amino acids and sequences of 50–2000 amino acids in different proteins. Each amino acid has its own shorthand code (e.g. ala). You do not need to know these.
- The amino acid sequence determines the properties of proteins and accounts for the diversity of proteins in living things. But each protein has its own unique primary structure.

SALTERS M4

Secondary structure

The **secondary** structure of a protein is a region in which there are regular structures.

There are two types, an **α-helix** and a **β-sheet**, each held together by **hydrogen bonds** between C=O and N–H:

C=O ···· H–N

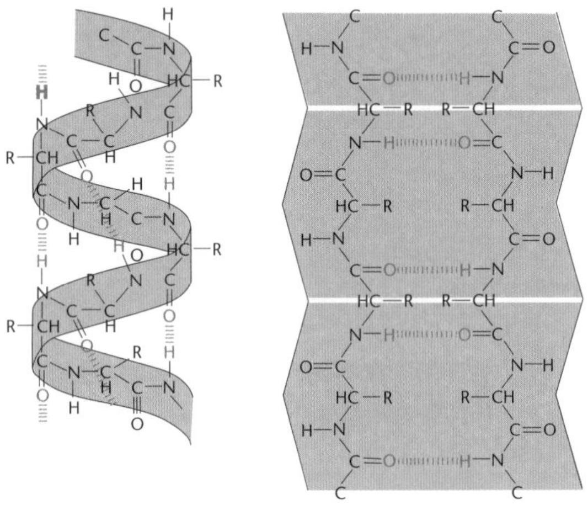

α-helix **β-sheet**

SALTERS M4

Tertiary structure

The **tertiary** structure of a protein is the **three-dimensional** shape. The tertiary structure will contain regions of secondary structures and will always have the sequence of amino acids in the primary structure.

The tertiary structure is held together by intermolecular bonds that act between R–groups of different amino acids.

Possible R– group interactions are:

The tertiary structure determines the properties of proteins.

- **hydrogen bonding** – between polar side chains such as CH_2OH and CH_2CONH_2
- **covalent bonding** caused by SH groups forming –S–S– linkages
- **ionic attractions** – between ions on side chains such as $–CH_2COO^-$ and NH_3^+
- **instantaneous dipole–induced dipole attractive forces** (van der Waals' forces) – between non-polar hydrocarbon chains.

Enzyme catalysis

SALTERS M4

Specificity

An **enzyme** is:

- a **biological catalyst**
- a protein with a definite **three-dimensional** tertiary structure.

An enzyme is **specific** about the reactions that can be catalysed. Within the tertiary structure, there is an **active site**.

The substrate has to have the correct shape but the intermolecular bonds must also fit. Otherwise there will be nothing to hold the substrate to the active site.

- The active site has a definite shape that will only accept a molecule if it is the correct shape – the '**Lock and Key**' analogy.
- The molecule that fits the active site is called the **substrate**.
- The substrate **bonds** to the active site by **intermolecular** bonds.

By bonding to the active site, a reaction is able to take place with a pathway of lower activation energy.

The effect of pH on enzyme activity

Enzymes work at a particular **optimum pH**. Changes in pH will affect the ability of the substrate to bond by intermolecular bonds to the active site.

Extremes of pH will interfere with the intermolecular bonds within the tertiary structure. This can change the shape of the active site so that the substrate no longer fits.

The effect of temperature on enzyme activity

Higher temperatures increase the **rate** of enzyme-catalysed reactions at the active site. However, if the **temperature** is **too high**,

- the **intermolecular bonds** within the **tertiary structure** may start to **break**,
- the **shape** of the **active site is changed** and the substrate no longer fits.

This change is irreversible and the enzyme will no longer function – it is **denatured**.

If you study Biology, you may have also come across non-competitive inhibition in which an inhibitor binds to another site on the enzyme, changing the shape of the active site.

Inhibition

An **inhibitor** is a molecule with a **similar shape** and with a similar arrangement of **binding sites** as the substrate.

- The inhibitor **competes** with the substrate for the active site and may block the site for access to a substrate molecule.
- The process is **reversible** as the inhibitor will also leave the active site.
- The rate of the enzyme-catalysed reaction slows down.

This is known as **competitive inhibition**.

Rates of enzyme-catalysed reactions

The graph below shows how the rate of an enzyme-catalysed reaction changes with increasing substrate concentration.

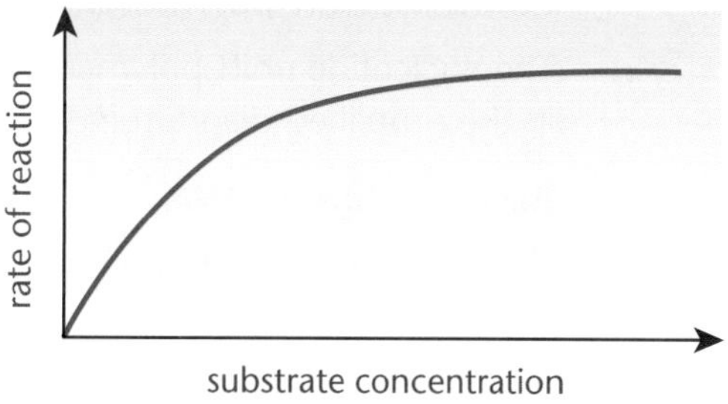

See page 17 for shapes of rate/concentration graphs for different orders of reaction.

At **low substrate concentrations**:

- the rate of reaction is **proportional** to the substrate concentration – the order with respect to substrate molecules is **one**
- active sites are **available** for any increase in substrate molecules.

As substrate **concentration increases**:

- there is progressively less effect until the rate levels off.
- the active sites become progressively **occupied**
- eventually all active sites are occupied and increasing the substrate concentration makes no difference to the rate – the order with respect to substrate molecules is **zero**.

Nucleic acids: DNA and RNA

SALTERS M4

DNA is a condensation polymer formed from monomer units called **nucleotides**, having three components: **phosphate**, **sugar** and **base**.

- The sugar is **deoxribose** in **DNA** (deoxyribonucleic acid).
- The sugar is **ribose** in **RNA** (ribonucleic acid).

Phosphate and sugars

Various models for DNA were devised before the currently accepted version was formulated.

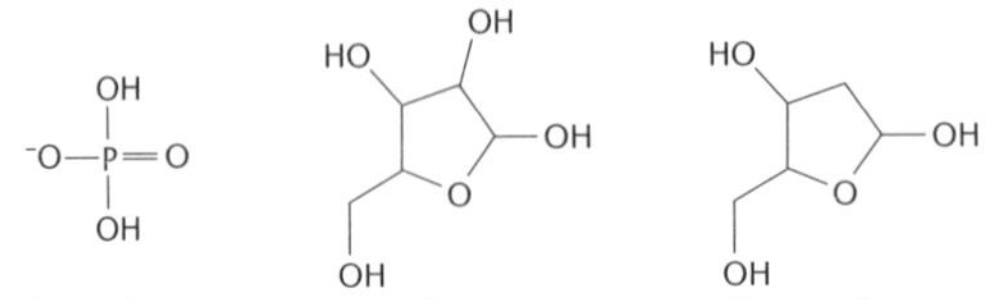

phosphate ribose deoxyribose

Bases

DNA has thymine.
RNA has uracil.
The other three bases are the same in DNA and RNA.

Each **nucleotide** can have one of **four** possible bases:

- DNA has: adenine; **thymine**; cytosine; guanine.
- RNA has: adenine; **uracil**; cytosine; guanine.

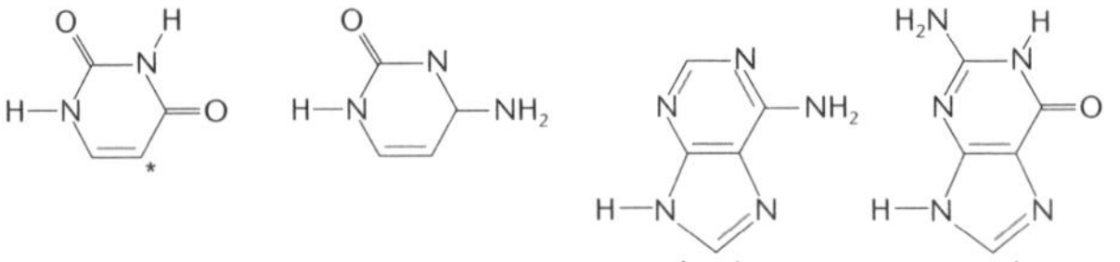

uracil cytosine adenine guanine

(thymine has a –CH_3 at position *)

The nucleotides

Each nucleotide has this sequence: **phosphate–deoxyribose–base**.
The phosphate unit and base are joined by condensation with deoxyribose, as shown below.

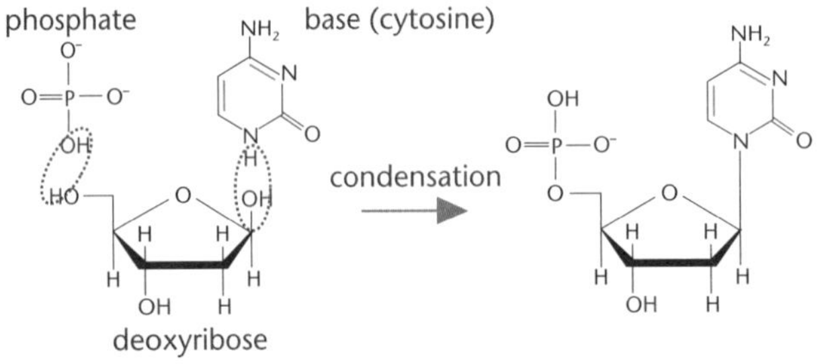

Condensation polymerisation of nucleotides

The nucleotide units polymerise through the sugar–phosphate backbone as shown in the simplified diagram in the margin.

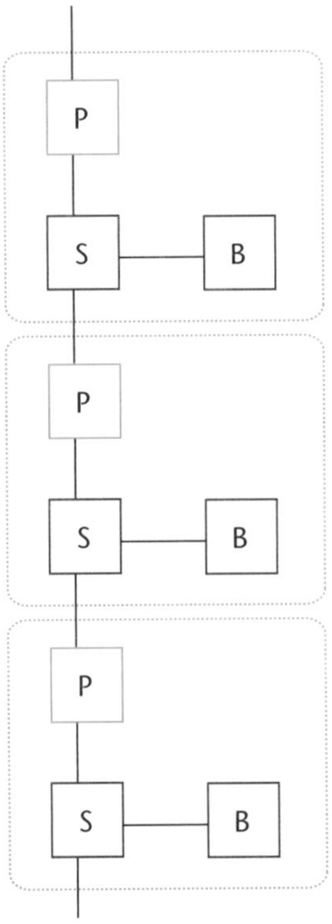

Simplified section of a DNA strand

- P = phosphate
- S = sugar
- B = base

Base pairing in DNA

Two separate **strands** are linked together in the **double helix** structure of DNA.
In DNA,

- **A**denine and **T**hymine pair up with **two** hydrogen bonds.
- **C**ytosine and **G**uanine pair up with **three** hydrogen bonds.

The geometry of each base ensures that the correct bases pair up.

- A and T 2 H bonds
- C and G 3 H bonds

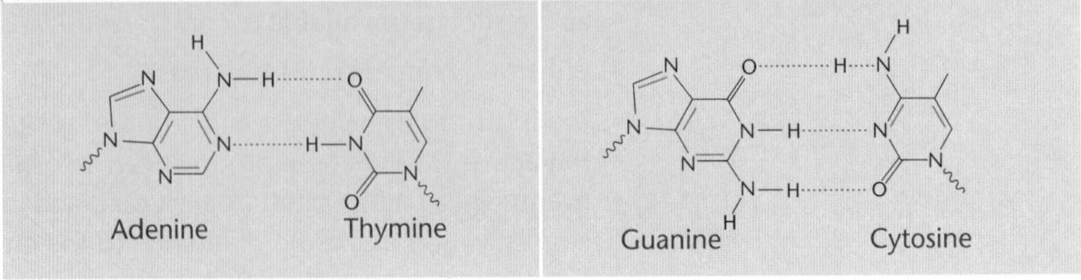

DNA analysis can be used for genetic fingerprinting. There are discussions about developing a national DNA database of every person in the UK but this carries with it ethical issues and concern about storage of personal information.

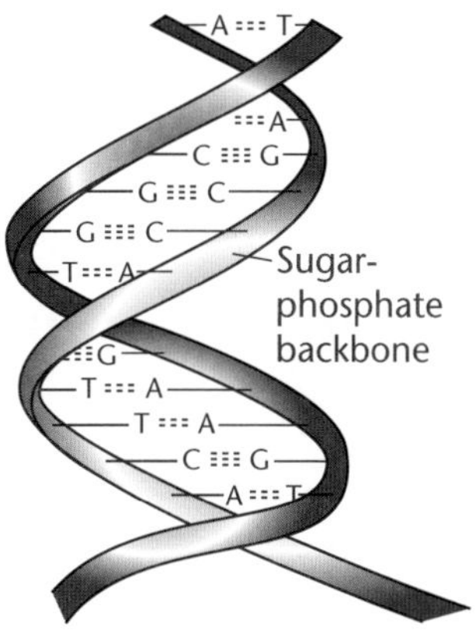

Each strand of DNA contains a sequence of bases in a definite order. The **complimentary strand** must contain only the correct sequence of the complimentary bases (see the margin diagram).

DNA can **replicate** by first 'unzipping' each strand:

- Each separate strand attracts nucleotides.
- By base pairing, the nucleotides are attached in the correct order.
- The result is **two** double helices of DNA, made from one original double helix.
- Each new double helix is identical to the original.

Every person's DNA is unique but sections of DNA are identifiable as **genes**. Some genes are inherited and are handed down through generations. These genes give family characteristics.There are also other genes that are common to every person's DNA and these carry the necessary information for the body to create proteins.

DNA and proteins

DNA encodes for an amino acid sequence in a protein by using a base sequence within a '**protein gene**' section of DNA.

The DNA double helix can make a strand of RNA by base pairing.

RNA has some important differences:

- A strand contains four bases but with the base **uracil** instead of **thymine**.
- **Ribose** is the sugar instead of deoxyribose.
- The strand is formed from just a section of the DNA double helix; it is much shorter than a DNA strand.
- RNA exists as a single strand.

The protein code

The strand of RNA formed is called **messenger RNA, m-RNA**.

- The sequence of bases in m-RNA contains **triplet codes**, each with **three** bases.
- Each amino acid can bond to another type of RNA strand called **transfer RNA, t-RNA** within the ribosome region of a cell.
- Each **amino acid** has its own **special t-RNA** with a special sequence of bases that have triplet codes – these are **complimentary** to the code in m-RNA.
- The **m-RNA** strand **attracts t-RNA** molecules in sequence, each carrying an amino acid. The triplet codes ensure that the correct t-RNA molecule is attached.

The diagram below shows a section of the m-RNA strand with 12 bases: **four triplet codes**.

Diagram to illustrate protein synthesis

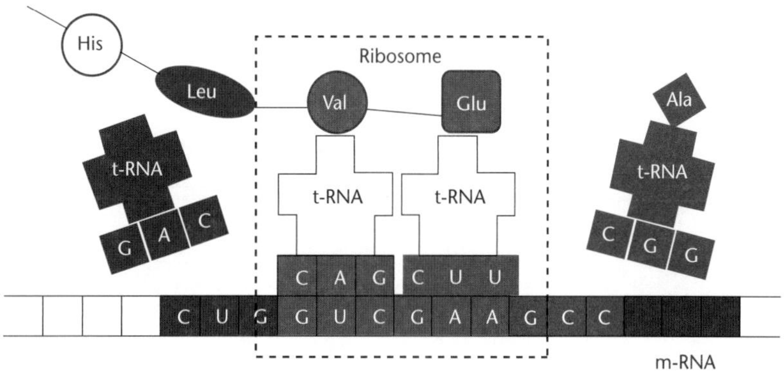

The amino acids join together in the correct sequence for a protein. As soon as two amino acid molecules join by a peptide link, the t-RNA molecule leaves to collect another amino acid molecule.

In the left-hand side of the diagram above, the t-RNA molecule that codes for Leu (leucine) is leaving. Note the correct base pairing within each matching of triplets.

Ala (alanine) is the next amino acid that needs to be added and you can see this on the right-hand side of the 'conveyor belt'.

5.3 Chromatography

After studying this section you should be able to:

- *understand what is meant by chromatography*
- *know that components are separated by absorption and partition*
- *understand what is meant by R_f value and retention time*
- *know that chromatography can be combined with mass spectrometry*

LEARNING SUMMARY

Types of chromatography

OCR M4
SALTERS M4, M5

Chromatography is an analytical technique that separates components in a mixture between a **mobile phase** and a **stationary phase**. The technique is especially useful for the purification of an organic substance.

- The **mobile phase** moves in a definite direction and may be a liquid (as in thin-layer chromatography, TLC) or a gas (as in gas chromatography, GC).
- The **stationary phase** is the substance that is fixed in place during the chromatography. It may be a solid (as in TLC) or either a liquid or solid on a solid support (as in GC).
- A solid stationary phase separates by **adsorption** of the components in the mixture on the surface.
- A liquid stationary phase separates by **partition** between the liquid and the mobile phase.

OCR M4
SALTERS M4

Thin-layer and paper chromatography

Thin-layer chromatography (TLC) uses a stationary phase in the form of a thin layer of an adsorbent such as silica gel or alumina on a flat, inert support.

Paper chromatography and TLC use a similar technique.

- A small dot of sample solution is placed on the TLC plate or paper.
- The paper or plate is placed in a jar containing a shallow layer of solvent and sealed.
- As the solvent rises, it meets the sample of the mixture. Different compounds in the mixture travel different distances depending on how strongly they interact with the paper and their solubility in the solvent.

This allows the calculation of a **Retardation factor,** R_f for each component (see below).

$$R_f = \frac{\text{distance moved by component}}{\text{distance moved by solvent front}}$$

KEY POINT

The R_f value can be compared to standard compounds to aid in the identification of an unknown substance. The diagram below shows some R_f values.

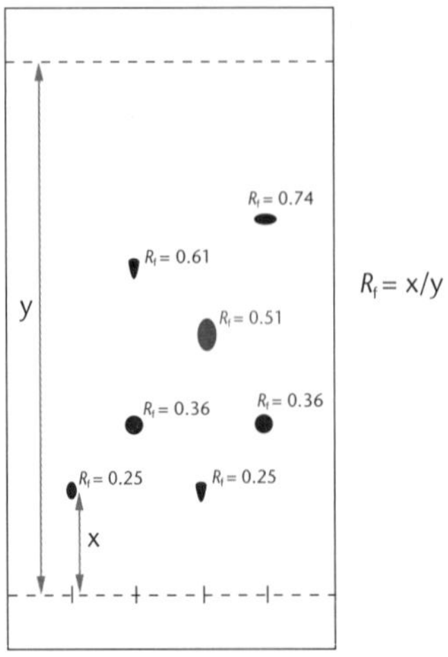

Using TLC, the separated components can be located as spots using iodine or ultraviolet radiation. Components can be identified by comparing the R_f values with known compounds.

The amino acids in a mixture can be separated and identified by paper chromatography and located using ninhydrin.

OCR M4
SALTERS M5

Gas Chromatography GC

Gas chromatography (GC) or Gas-Liquid chromatography (GLC) is a separation technique in which the mobile phase is a gas. Gas chromatography is always carried out in a column. The column contains the stationary phase, which is either a high-boiling liquid or a solid on a solid, porous support.

- The sample is injected into the column, which is heated to vaporise the components in the mixture.
- The gas mobile phase flushes the mixture along the column.
- As the mixture moves through the column, components are slowed down as they interact with the stationary phase. Different components are slowed down by different amounts, which causes them to separate.
- Each component leaves the column at a different time.

This allows the calculation of the retention time for each component.

> **Retention time** is the time it takes for a component to pass from the column inlet to the detector.
>
> KEY POINT

The retention time can be compared to standard compounds to aid in the identification of an unknown substance. The peak area also gives the approximate percentage composition of a mixture. The diagram below shows a GC chromatogram with some retention times.

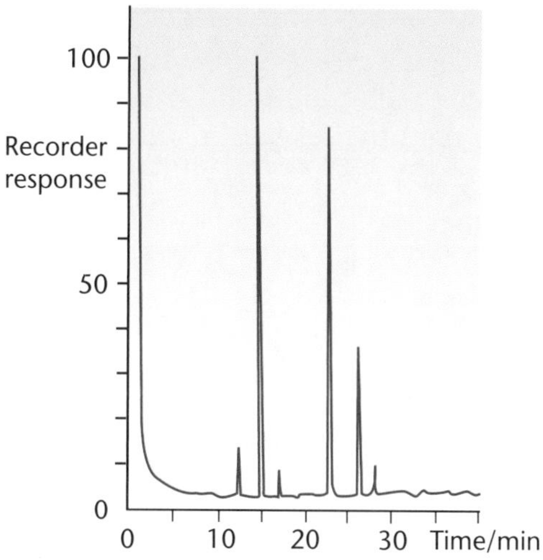

Limitations

Analysis by gas chromatography has limitations:

- Similar compounds often have similar retention times.
- Unknown compounds have no reference retention times for comparison.

Combining mass spectrometry with chromatography

Mass spectrometry can be combined with chromatography to provide a far more powerful analytical tool than chromatography alone. For example, mass spectrometry is used with GC as GC-MS and with high pressure liquid chromatography as HPLC-MS.

GC-MS is extensively used in modern analysis, e.g. in forensics, environmental analysis and airport security.

In GC-MS:

- the components in a mixture are first separated using GC
- the separated components are directed into the mass spectrometer
- the generated mass spectra can be analysed or compared with a spectral database by computer for positive identification of each component.

Progress check

1 What is meant by the following terms?
(a) R_f value
(b) retention time

2 How does GC-MS enable components in a mixture to be identified?

1 (a) $R_f = \frac{\text{distance moved by component}}{\text{distance moved by solvent front}}$
(b) The time it takes for a component to pass from the column inlet to the detector.

2 GC separates the components. Each component then passes into a mass spectrometer. The resulting mass spectrum is analysed against a spectral database to identify the component.

5.4 Analysis

After studying this section you should be able to:

- *review from AS Chemistry the use of infra-red spectroscopy and mass spectrometry in organic analysis*

LEARNING SUMMARY

Infra-red spectroscopy and mass spectrometry in organic analysis

OCR M4
SALTERS M4, M5

Key points from AS

- **Analysis**
  Revise AS pages 119–122

In AS Chemistry, you were introduced to the use of two important instrumentation techniques in organic analysis:

- Infra-red spectroscopy.
- Mass spectrometry.

For A2 chemistry, you study a further important instrumentation technique called nuclear magnetic resonance, NMR. Analytical chemists often combine the best features of each spectral technique to identify an unknown compound. This section starts with a review of the key features of IR and MS. For full details, you will need to refer to *Revise AS Chemistry*, Chapter 4.7.

Using infra-red (IR) spectroscopy in organic analysis

IR spectroscopy is useful for identifying the functional groups in a molecule.

An infra-red spectrum is particularly useful for identifying:

- an alcohol from absorption of the O–H bond (3230–3550 cm^{-1})
- a carbonyl compound from absorption of the C=O bond (1680–1750 cm^{-1})
- a carboxylic acid from absorption of the C=O bond (1680–1750 cm^{-1}) and broad absorption of the O–H bond (2500–3300 cm^{-1} (broad)).

IR spectroscopy is most useful for identifying C=O and O–H bonds.

Look for the distinctive patterns.

Fingerprint region

- Between 1000 and 1550 cm^{-1}

Many spectra show a complex pattern of absorption in this range.

- This pattern can allow the compound to be identified by comparing with spectra of known compounds.

The fingerprint region is unique for a particular compound.

Using mass spectrometry in organic analysis

Mass spectrometry of organic compounds is useful for identifying:

- the relative molecular mass from the molecular ion peak, *M*
- parts of the skeletal formula by the fragmentation pattern.

The fragmentation pattern provides clues about the molecular structure of the compound. The table below shows the identity of the main peaks in the mass spectrum of butanone, $CH_3COCH_2CH_3$.

m/z	*ion*	*fragment lost*
72	$CH_3COCH_2CH_3^+$	–
57	$CH_3CH_2CO^+$	CH_3•
43	CH_3CO^+	CH_3CH_2•
29	$CH_3CH_2^+$	CH_3CO•
15	CH_3^+	CH_3CH_2CO•

Note that only ions are detected in the mass spectrum.

Uncharged species such as the radical cannot be deflected within the mass spectrometer.

The skill in interpreting a mass spectrum is in searching for known patterns, evaluating all the evidence and reassembling all the information into a molecular structure.

Using infra-red (IR) spectroscopy in analysis

OCR M4
SALTERS M4, M5

Basic principles

Bonds in molecules naturally vibrate. Some bonds in molecules increase their vibrations by absorbing energy from IR radiation. Different bonds absorb different frequencies of IR radiation.

An IR spectrum is obtained by passing a range of IR frequencies through a compound. As energy is taken in, **absorption peaks** are produced. The frequencies of the absorption peaks can be matched to those of known bonds to identify structural features in an unknown compound.

The frequency of IR absorption is measured in wavenumbers, units: cm^{-1}.

IR radiation has less energy than visible light.

KEY POINT

IR spectroscopy is useful for identifying the functional groups in a molecule.

Important IR absorptions

bond	*functional group*	*wavenumber/cm⁻¹*
O–H	hydrogen bonded in alcohols	3230 – 3550
N–H	amines	3100 – 3500
C–H	organic compound with a C–H bond	2850 – 3100
O–H	hydrogen bonded in carboxylic acids	2500 – 3300 (broad)
C≡N	nitriles	2200 – 2260
C=O	aldehydes, ketones, carboxylic acids, esters	1680 – 1750
C–O	alcohols, esters	1000 – 1300

You don't need to learn the absorption frequencies – the data is provided.

IR spectroscopy is used in some modern breathalysers for measuring the concentration of blood alcohol. A particular IR absorption identifies the presence of ethanol in the breath and the intensity of the peak is directly related to the ethanol level.

An infra-red spectrum is particularly useful for identifying:

- an **alcohol** from absorption of the O–H bond
- a **carbonyl** compound from absorption of the C=O bond
- a **carboxylic acid** from absorption of the C=O bond **and** broad absorption of the O–H bond.

Interpreting infra-red spectra

OCR M4
SALTERS M4, M5

Carbonyl compounds (aldehydes and ketones)
Butanone, $CH_3COCH_2CH_3$

- C=O absorption 1680 to 1750 cm^{-1}

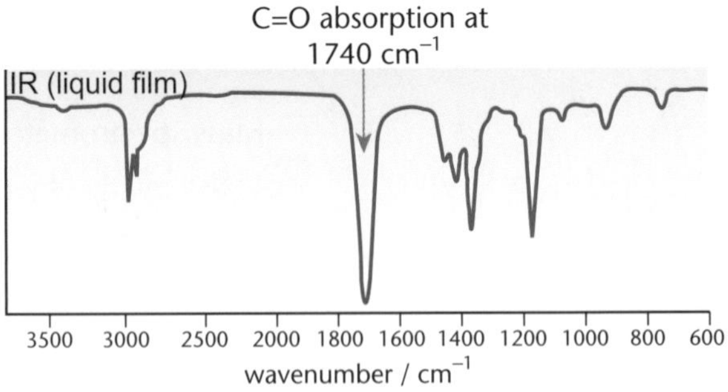

Alcohols
Ethanol, C_2H_5OH

- O–H absorption 3230 to 3500 cm^{-1}
- C–O absorption 1000 to 1300 cm^{-1}

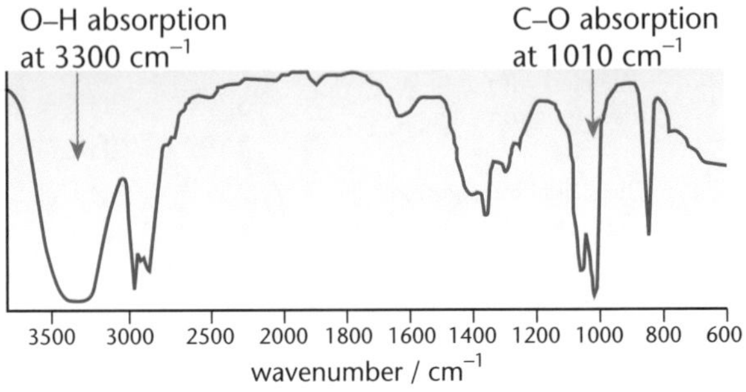

IR spectroscopy is most useful for identifying C=O and O–H bonds.

Look for the distinctive patterns.

Note that all these molecules contain C–H bonds, which absorb in the range 2840 to 3045 cm^{-1}.

Carboxylic acids

Propanoic acid,
C_2H_5COOH

- Very broad O–H absorption 2500 to 3500 cm^{-1}
- C=O absorption 1680 to 1750 cm^{-1}

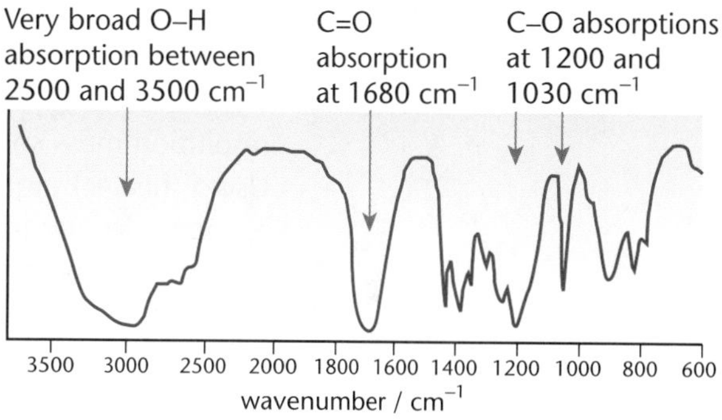

There are other organic groups (e.g. N–H, C=C) that absorb IR radiation but the principle of linking the group to the absorption wavenumber is the same.

Ester

Ethyl ethanoate,
$CH_3COOC_2H_5$

- C=O absorption 1680 to 1750 cm^{-1}
- C–O absorption 1000 to 1300 cm^{-1}

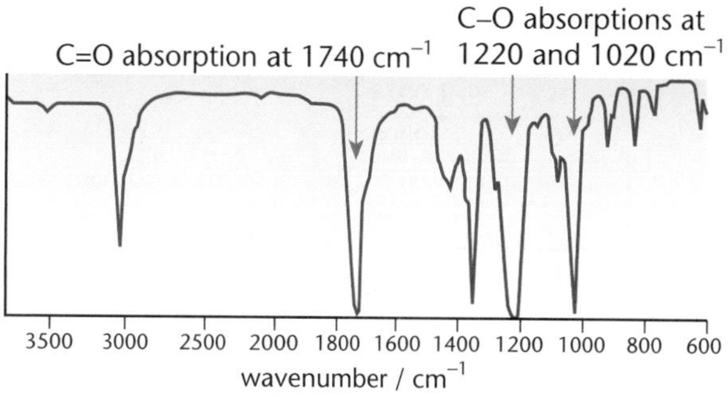

The fingerprint region is unique for a particular compound.

Fingerprint region

- Between 1000 and 1550 cm^{-1}

Many spectra show a complex pattern of absorption in this range.

- This pattern can allow the compound to be identified by comparing its spectrum with spectra of known compounds.

Mass spectrometry

OCR	M4
SALTERS	M4, MS

Mass spectrometry can be used to determine relative atomic masses from a mass spectrum. Mass spectrometry is also used to determine relative molecular masses and to identify the molecular structures of organic compounds.

Molecular ions

The molecular ion peak is usually given the symbol M.

The mass spectrum also contains fragments of the molecule ion (see detail below).

Organic molecules can be analysed using mass spectrometry.

In the mass spectrometer, organic molecules are bombarded with electrons. This can lead to the formation a **molecular ion**.

The equation below shows the formation of a molecular ion from butanone, $CH_3COCH_2CH_3$.

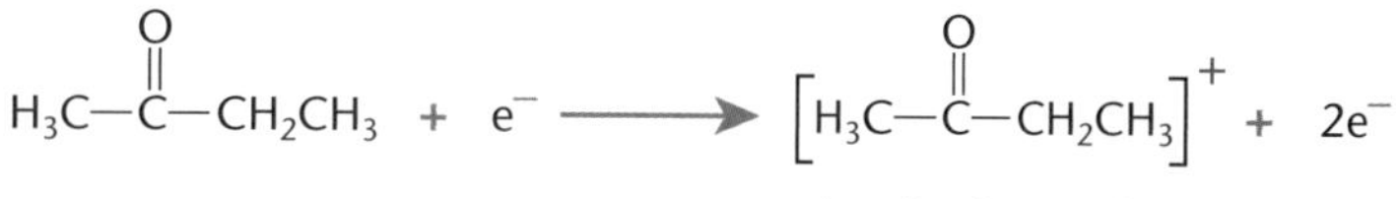

molecular ion, *m/z*: 72

- The molecular ion peak, M, provides the relative molecular mass of the compound.

High resolution mass spectrometry

SALTERS M4

Molecules of similar relative molecular masses can be distinguished using high resolution mass spectrometry.

Using this technique:

- the molecular mass is obtained with a very accurate value
- this is compared with very accurate relative isotopic masses.

isotope	*relative isotopic mass*
^{1}H	1.0078
^{12}C	12.0000
^{14}N	14.0031
^{16}O	15.9949

Worked Example

Three gases, CO, C_2H_4 and N_2, each have an approximate relative molecular mass of 28.

The table below shows accurate molecular masses for these gases, calculated from accurate relative isotopic masses (see the margin).

gas	*CO*	*C_2H_4*	*N_2*
molecular mass	27.9949	28.0312	28.0062

The gases can be distinguished by matching the *m/z* values from high resolution mass spectrometry with the calculated values.

Fragment ions

OCR M4

SALTERS M4, M5

In the conditions within the mass spectrometer, some molecular ions are fragmented by bond fission.

- The bond fission that takes place is a fairly random process:
  - different bonds are broken
  - a **mixture** of **fragment ions** is obtained.
- The mass spectrum contains both the molecular ion and the mixture of fragment ions.

The fragmentation pattern provides clues about the molecular structure of the compound.

Mass spectrometry of organic compounds is useful for identifying:

- the relative molecular mass
- parts of the skeletal formula.

The mass spectrum of butane would contain peaks for these four ions:

$C_4H_{10}^{+}$: $m/z = 58$

$C_3H_7^{+}$: $m/z = 43$

$C_2H_5^{+}$: $m/z = 29$

CH_3^{+}: $m/z = 15$

Fragmentation of butane, C_4H_{10}

Bond fission forms a fragment ion and a radical. The diagram below shows that fission of the same C–C bond in a butane molecule can form two different fragment ions.

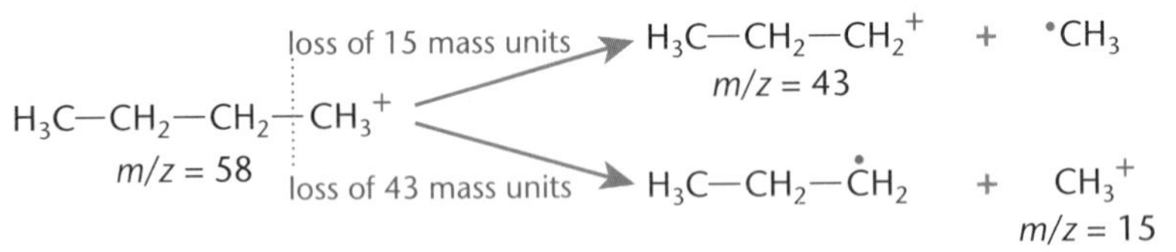

- Fragmentation of the molecular ion produces a fragment ion by loss of a radical.
- The mass spectrometer only detects the fragment ion.
- The fragment ion may itself fragment into another ion and radical.

Note that only ions are detected in the mass spectrum.

Uncharged species such as the radical cannot be deflected within the mass spectrometer.

The mass spectrum of butanone

The mass spectrum of butanone, $CH_3CH_2COCH_3$, is shown below. The *m/z* values of the main peaks have been labelled.

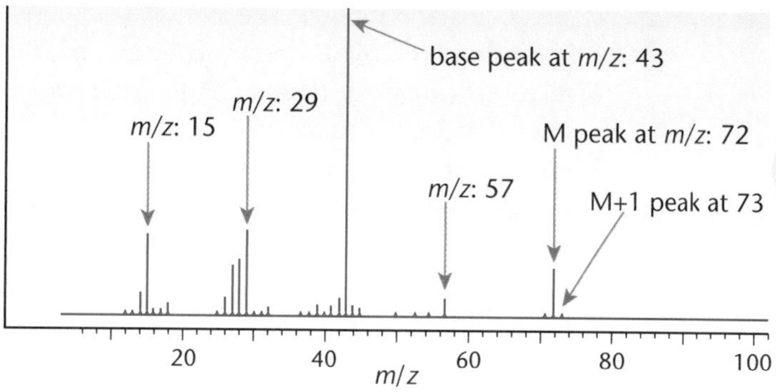

The M+1 peak is a small peak 1 unit higher than the molecular ion peak.

The origin of the M+1 peak is the small proportion of carbon-13 in the carbon atoms of organic molecules.

The table below shows the identities of the main peaks.

A common mistake in exams is to show both the fragmentation products as ions.

m/z	*ion*	*fragment lost*
72	$CH_3COCH_2CH_3^+$	–
57	$CH_3CH_2CO^+$	CH_3•
43	CH_3CO^+	CH_3CH_2•
29	$CH_3CH_2^+$	CH_3CO•
15	CH_3^+	CH_3CH_2CO•

The acylium (RCO^+) ion is particularly stable and is often present as a high dominant peak.

The base peak is given a relative abundance of 100. Other peaks are compared with this base peak.

The most abundant peak in the mass spectrum is the base peak.

- The base peak in the mass spectrum of butanone has *m/z*: 43, formed by the loss of a $CH_3CH_2^•$ radical:

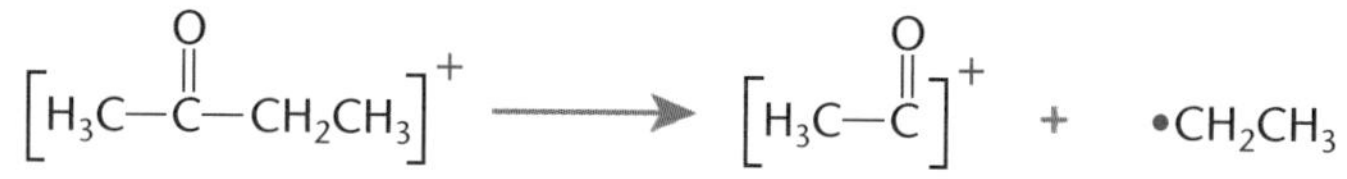

Common patterns in mass spectra

Different fragmentations are possible depending on the structure of the molecular ion.

- The table above shows common fragment ions and fragmented radicals.
- The skill in interpreting a mass spectrum is in searching for known patterns, evaluating all the evidence and reassembling all the information into a molecular structure.
- You should also look for a peak at *m/z* 77 or loss of 77 units. This is a giveaway for the presence of a phenyl group, C_6H_5, in a molecule.

Progress check

1 Write an equation for the formation of the fragment ion at *m/z*=29 in the mass spectrum of butanone above.

2 What peaks would you expect in the mass spectrum of pentane?

1 $CH_3CH_2COCH_3^+ \longrightarrow CH_3CH_2^+ + CH_3CO$•

2 *m/z*: 72, $C_5H_{12}^+$; *m/z*: 57, $C_4H_9^+$; *m/z*: 43, $C_3H_7^+$; *m/z*: 29, $C_2H_5^+$; *m/z* : CH_3^+, 15.

5.5 NMR spectroscopy

After studying this section you should be able to:

LEARNING SUMMARY

- *understand that NMR spectroscopy is carried out with ^{1}H and ^{13}C*
- *predict the different types of proton and carbon present in a molecule from chemical shift values*
- *predict the relative numbers of each type of proton present from an integration trace*
- *predict the number of protons adjacent to a given proton from the spin-spin coupling pattern*
- *predict possible structures for a molecule from ^{1}H and ^{13}C spectra*
- *predict the chemical shifts and splitting patterns of the protons in a given molecule*
- *describe the use of D_2O in NMR spectroscopy*

Nuclear magnetic resonance

OCR M4
SALTERS M5

Nuclear Magnetic Resonance (NMR) spectroscopy is an extremely important modern method of analysis. It is used extensively in the pharmaceutical industry and in universities for identification and purity checking of organic compounds.

Key principles

Only nuclei with an odd number of nucleons (neutrons + protons) possess a magnetic spin: e.g. ^{1}H, ^{13}C. Proton NMR spectroscopy is the most useful general purpose technique.

The nucleus of an atom of hydrogen (i.e. a proton) has a magnetic spin.
When placed in a strong electromagnetic field:

- the nucleus can absorb energy from the low energy **radio-frequency** region of the spectrum to move to a higher energy state
- **nuclear magnetic resonance** occurs as protons resonate between their spin energy states.

The energy gap is equal to that provided by radio waves.

The different nuclear spin states in an applied magnetic field

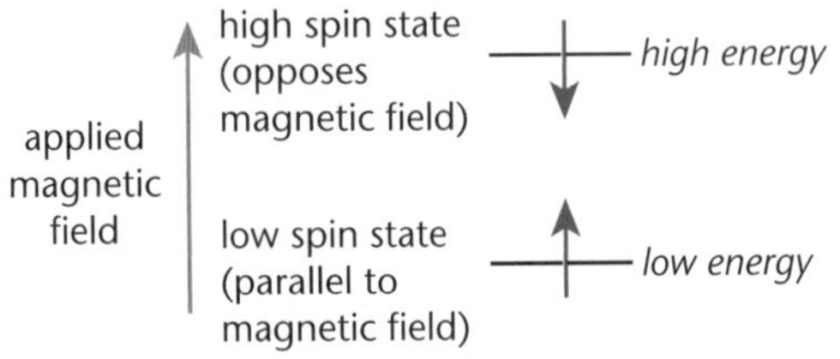

An NMR spectrum shows **absorption peaks** corresponding to the radio-frequency absorbed.

Chemical shift, δ

Electrons around the nucleus **shield** the nucleus from the applied field.

- The magnetic field at the nucleus of a particular proton is different from the applied magnetic field.

NMR spectroscopy is the same technology as used in MRI (magnetic resonance imaging) in medical body scanners. It is a non-invasive technique.

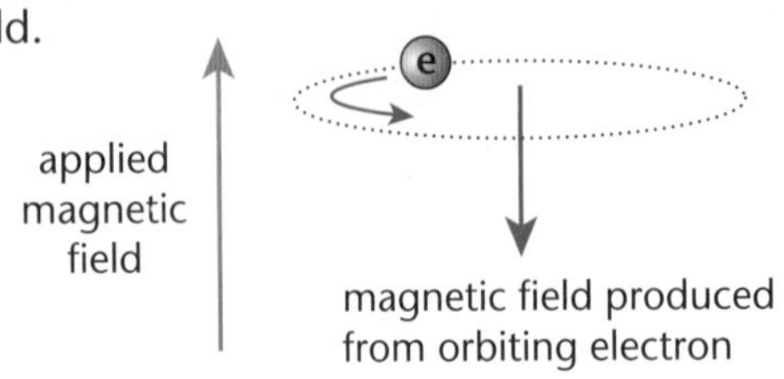

Protons in different environments absorb at different chemical shifts.

- Different radio-frequencies are absorbed, depending on the **environment** of the proton.
- **Chemical shift** is a measure of the magnetic field experienced by protons in different environments resulting from nuclear shielding.

> Chemical shift, δ, is measured relative to a standard: tetramethylsilane (TMS), $Si(CH_3)_4$.
> • The chemical shift of TMS is defined as δ = 0 ppm.
>
> KEY POINT

Carbon-13 NMR spectroscopy

OCR M4

Carbon-13 NMR spectroscopy allows the identification of carbon atoms in an organic molecule and is an important tool in the determination of structure in organic chemistry.

Typical chemical shifts

Chemical shifts indicate the chemical environment of the **carbon atoms** present. The table below shows typical chemical shift values for different types of carbon atom.

You don't need to learn these chemical shifts - the data is provided on exam papers.

The presence of an electronegative atom or group causes chemical shift 'downfield'. This is called 'deshielding'.

Notice the chemical shift caused by the carbonyl group, oxygen, halogens and a benzene ring.

type of carbon	*chemical shift, δ/ppm*
C–C	5–55
C–Cl or C–Br	30–70
C–N (amines)	35–60
C–O	50–70
C=C (alkenes)	115–140
aromatic 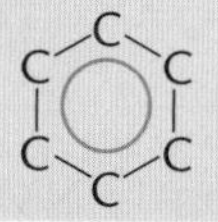	110–165
carbonyl (ester, carboxylic acid, amide) 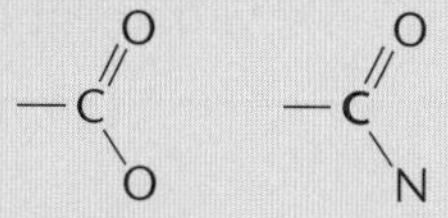	160–185
carbonyl (aldehyde, ketone) 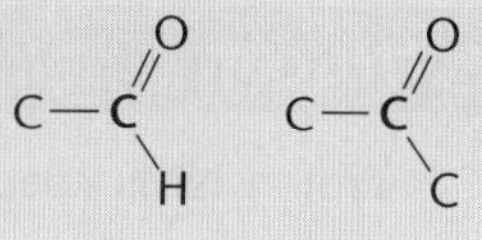	190–220

- The actual chemical shift may be slightly different depending upon the actual environment of the carbon.

In a carbon–13 NMR spectrum, the size of the peak tells us nothing about the number of carbon atoms responsible. This is in contrast to proton NMR spectroscopy. However, carbon-13 spectrum is far more sensitive with a much larger range of chemical shift values. Unless there are carbon atoms present in an identical environment, each carbon atom is likely to show as a separate single signal.

Interpreting a carbon-13 NMR spectrum

A carbon-13 NMR spectrum of ethyl ethanoate, $CH_3COOCH_2CH_3$ shows absorptions at **four** chemical shifts showing carbon atoms in **four different environments**:

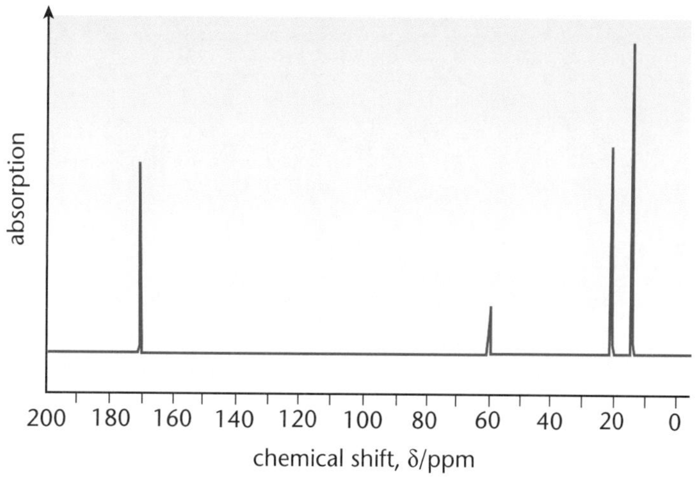

- The four chemical shifts can be matched to the table of chemical shifts on page 167 so that the carbon atom responsible for each absorption can be identified.

chemical shift/ppm	*environment*
171	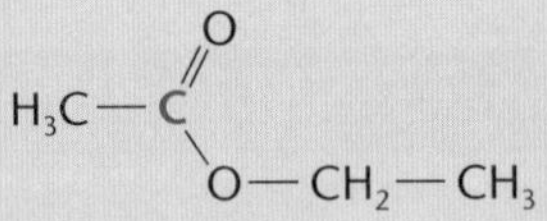
60	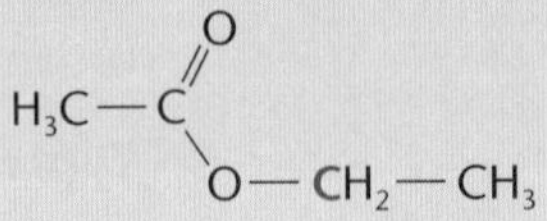
21	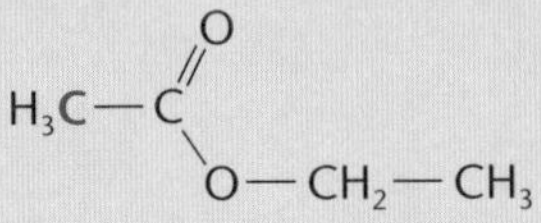
14	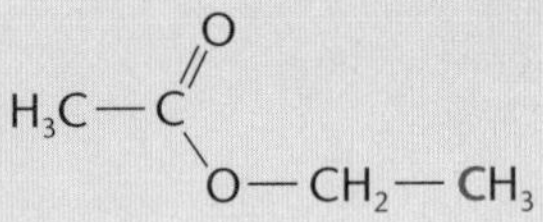

> **KEY POINT**
>
> Carbon-13 NMR spectroscopy is useful for identifying the **type** of environment of each carbon – the chemical shift.

Proton NMR spectroscopy

OCR M4
SALTERS M5

Proton NMR spectroscopy allows the identification of hydrogen atoms in an organic molecule and is an important tool in the determination of structure in organic chemistry.

Typical chemical shifts

Chemical shifts indicate the **types of protons** and **functional groups** present. The table on page 151 shows typical chemical shift values for protons in different chemical environments.

The presence of an electronegative atom or group causes chemical shift 'downfield'. This is called 'deshielding'.

Notice the chemical shift caused by the carbonyl group, oxygen, halogens and a benzene ring.

You don't need to learn these chemical shifts – the data is provided on exam papers.

type of proton	*chemical shift, δ/ppm*
R-CH_3	0.7–1.6
R-CH_2-R	1.2–1.4
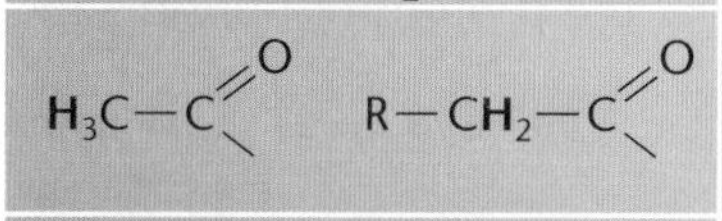	2.0–2.9
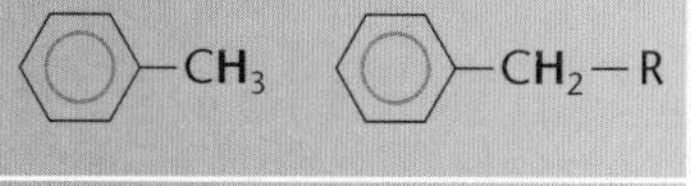	2.3–2.7
X-CH_3 X-CH_2-R (X = halogen)	3.2–3.7
-O-CH_3 -O-CH_2-R	3.3–4.3
R-O-**H**	3.5–5.5
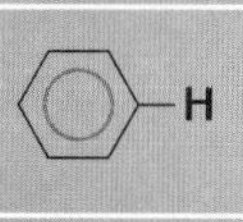	7
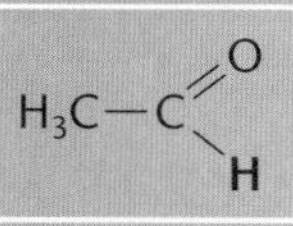	9.5–10
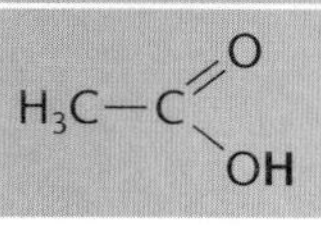	11.0–11.7

The only reliable means of identifying O–H in NMR is to use D_2O: see Identifying O–H protons, page 153.

- The actual chemical shift may be slightly different depending upon the actual environment of the proton.
- The chemical shift for O–**H** can vary considerably and depends upon concentration, solvent and other factors.

Progress check

1 For each structure, predict the chemical shift of each carbon atom.
(a) CH_3CH_2OH; (b) CH_3CH_2CHO; (c) CH_3COCH_3.

1 (a) CH_3CH_2OH, δ = 5–55 ppm; CH_3CH_2OH, δ = 50–70 ppm
(b) CH_3CH_2CHO, δ = 5–55 ppm; CH_3CH_2CHO, δ = 5–55 ppm; CH_3CH_2CHO, δ = 190–220 ppm
(c) CH_3COCH_3, δ = 5–55 ppm; CH_3COCH_3, δ = 190–220 ppm

Interpreting low resolution NMR spectra

OCR M4
SALTERS M5

Low resolution NMR spectrum of ethanol

A low resolution NMR spectrum of ethanol shows absorptions at **three** chemical shifts, showing the **three different types of proton**:

The relative areas of each peak are usually measured by running a second NMR spectrum as an *integration trace.*

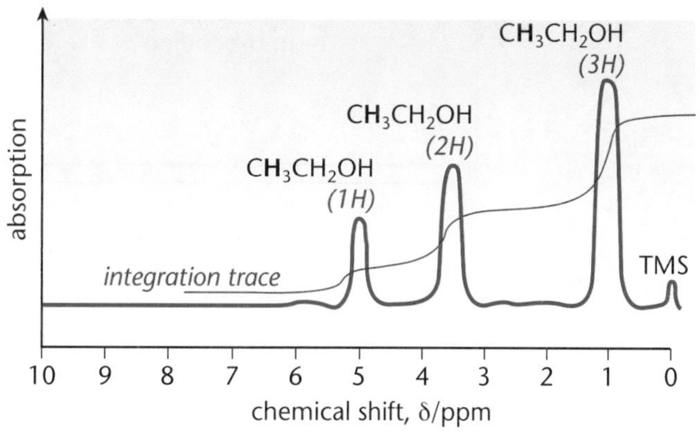

An NMR spectrum is obtained in solution. The solvent must be proton-free, usually CCl_4 or $CDCl_3$ is used.

- The **area** under each peak is in direct proportion to the **number of protons** responsible for the absorption.
- The three chemical shifts can be matched to the table of chemical shifts on page 151 so that the protons responsible for each absorption can be identified.

chemical shift/ppm	*no. of protons*	*environment*	*type of proton*
δ = 1.0	3H	$\mathbf{CH_3}CH_2OH$	**CH_3** adjacent to a carbon chain
δ = 3.5	2H	$CH_3\mathbf{CH_2}OH$	**CH_2** adjacent to –O
δ = 4.9	1H	$CH_3CH_2\mathbf{OH}$	**OH**

> **KEY POINT**
> A low resolution NMR spectrum is useful for identifying:
> - the **number** of different types of proton from the number of peaks
> - the **type** of environment of each proton from the chemical shift
> - **how many** protons of each type from the integration trace.

Interpreting high resolution NMR spectra

OCR M4
SALTERS M5

A high resolution NMR spectrum shows splitting of peaks into a pattern of sub-peaks.

Different types of proton have different chemical shifts.

Spin-spin coupling patterns:
- arise from interactions between protons on **adjacent** carbon atoms which have **different chemical shifts**
- indicate the number of **adjacent** protons.

Equivalent protons (i.e. protons with the same chemical shift) will **not** couple with one another.

The **spin-spin coupling** pattern shows as a **multiplet** – a doublet, triplet, quartet, etc.

A singlet is next to C.
A doublet is next to CH.
A triplet is next to CH_2.
A quadruplet is next to CH_3.

> **KEY POINT**
> We can interpret the spin-spin coupling pattern using the **n+1 rule.**
> For **n adjacent protons,** the number of peaks in a multiplet = **n+1.**

The high resolution NMR spectrum of ethanol

The high resolution NMR spectrum of ethanol shows spin-spin coupling patterns – some of the signals have been split into multiplets:

The multiplicity (doublet, triplet, quartet) does not indicate the number of protons on that carbon. The number of protons is given by the integration trace.

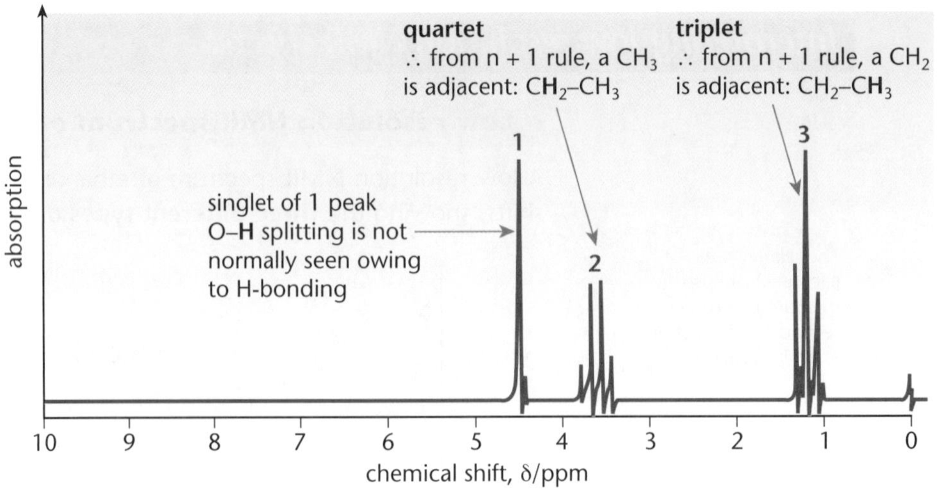

Notice the different sizes of each sub-peak within the splitting pattern:
- doublet 1:1
- triplet 1:2:1
- quartet 1:3:3:1.

- The **chemical shift** identifies the **type of protons** responsible for each peak.
- The **n+1** rule can be used to identify the **number of adjacent protons.**

chemical shift /ppm	*environment*	*number of adjacent protons (n)*	*splitting pattern (n+1)*
δ = 1.2	$\mathbf{CH_3}CH_2OH$	2H	2+1 = 3: **triplet**
δ = 3.6	$CH_3\mathbf{CH_2}OH$	3H	3+1 = 4: **quartet**
δ = 4.5	$CH_3CH_2O\mathbf{H}$	–	**singlet**

- Note that O–**H** splitting is not normally seen owing to H-bonding or exchange with the solvent used.

Equivalent protons do not interact with each other.

- The three equivalent CH_3 protons in ethanol cause splitting of the adjacent CH_2 protons, but not amongst themselves.

> A high resolution NMR spectrum is useful for identifying the **number** of **adjacent** protons from the spin-spin coupling pattern.
>
> KEY POINT

Identifying O–H protons

The OH absorption peak at 4.5 δ is absent in D_2O.

O–H protons can absorb at different chemical shifts depending on the solvent used and the concentration. The signals are often broad and usually show no splitting pattern.

These factors can make it difficult to identify O–H protons. However, they can be identified by using deuterium oxide, D_2O.

- A second NMR spectrum is run with a small quantity of D_2O added to the solvent.
- The D_2O exchanges with O–H protons and any such peak will disappear.

Solvents

Deuterated solvents such as $CDCl_3$ are used in NMR spectroscopy. Any protons in a solvent such as $CHCl_3$ would produce a large proton absorption peak. By using $CDCl_3$, this absorption is absent and the spectrum is that of the organic compound alone.

Progress check

1 For each structure, predict the number of peaks in its low resolution NMR spectrum corresponding to the different types of proton and also the number of each different type of proton
(e.g. CH_3CH_2OH has 3 peaks in the ratio 3:2:1).
(a) CH_3OH (b) CH_3CH_2CHO (c) CH_3COCH_3 (d) $(CH_3)_2CHOH$.

2 The NMR spectrum of compound **X** (C_2H_4O) has a doublet at δ 2.1 and a quartet at δ 9.8.
(a) Identify compound **X**. (Use the table of chemical shifts on page 151).
(b) How many protons are responsible for each multiplet?

1 (a) 2 peaks in the ratio 3:1 (b) 3 peaks in the ratio 3:2:1 (c) 1 peak
(d) 3 peaks in the ratio 6:1:1
2 (a) ethanal, CH_3CHO.
(b) doublet at δ 2.1, 3H; quartet at δ 9.8, 1H.

5.6 Organic synthetic routes

There are many variations possible and the schemes below could not include all reactions without appearing more complicated.

Organic chemists are frequently required to synthesise an organic compound in a multistage process. This is fundamental to the design of new organic compounds such as those needed for modern drugs to combat disease, a new dye or a new fibre. With so many reactions to consider, it is essential to see how different functional groups can be interconverted and summary charts are useful for showing these links. The flow-charts on the next pages show reactions drawn from both the AS and A2 parts of A Level Chemistry.

Construct a scheme of your own using only those reactions in your course.

In revision it is useful to draw your own schemes from memory.

Aliphatic synthetic routes

OCR M4
SALTERS M4, M5

The scheme below is based upon two main sets of reactions:

- a set based around bromoalkanes
- a set based around carboxylic acids and their derivatives.

The two sets of reactions are linked via the oxidation of primary alcohols.

- A reaction scheme based upon a secondary alcohol would result in oxidation to a ketone only.

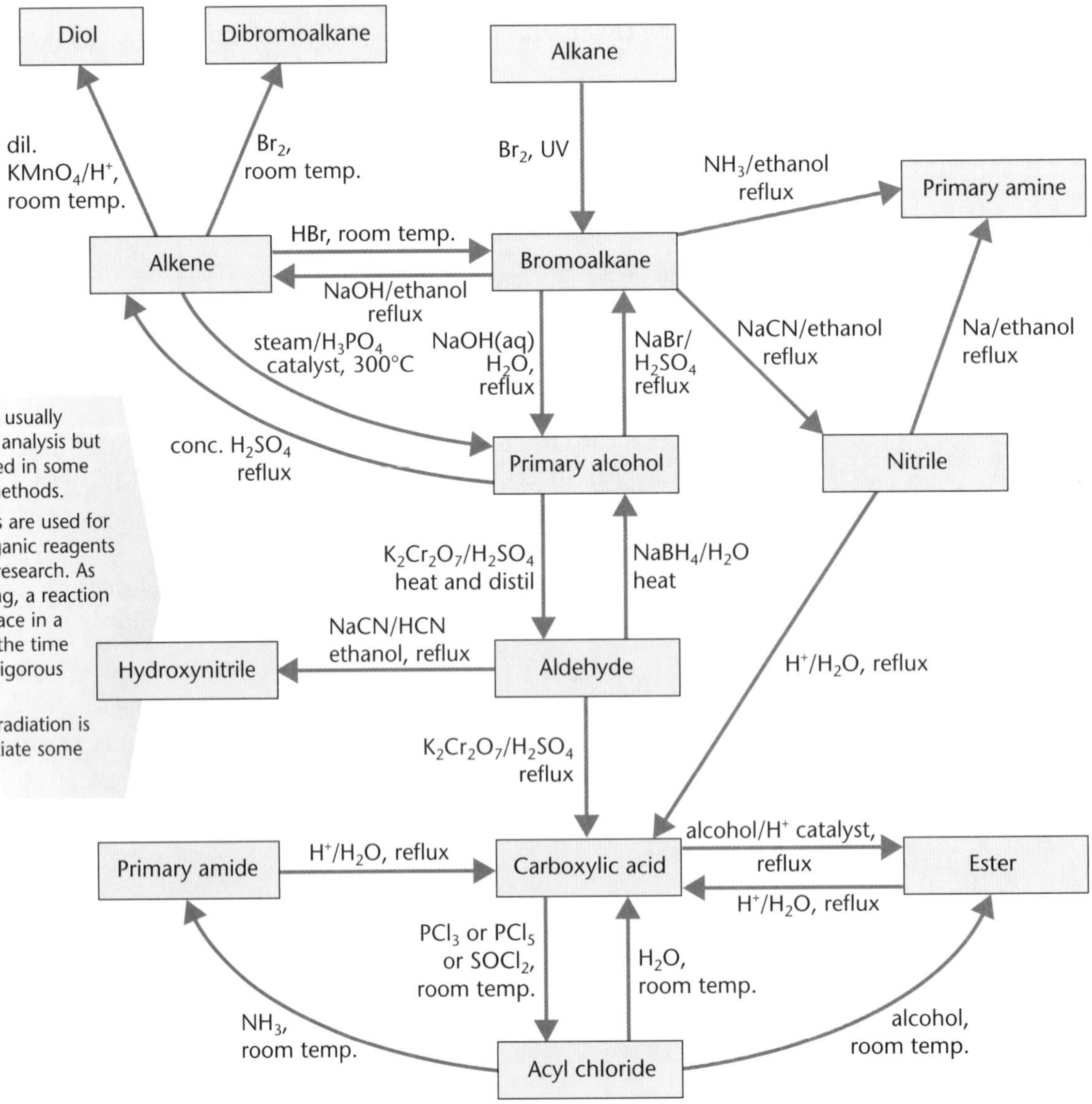

Radiation is usually linked with analysis but it is also used in some synthetic methods.

Microwaves are used for heating organic reagents in organic research. As with cooking, a reaction can take place in a fraction of the time using less vigorous conditions.

Ultraviolet radiation is used to initiate some reactions.

Aromatic synthetic routes

OCR M4
SALTERS M5

Compared with aliphatic organic chemistry at A Level, there are comparatively few aromatic reactions.

Benzene

- Reactions involving the benzene ring are mainly electrophilic substitution.

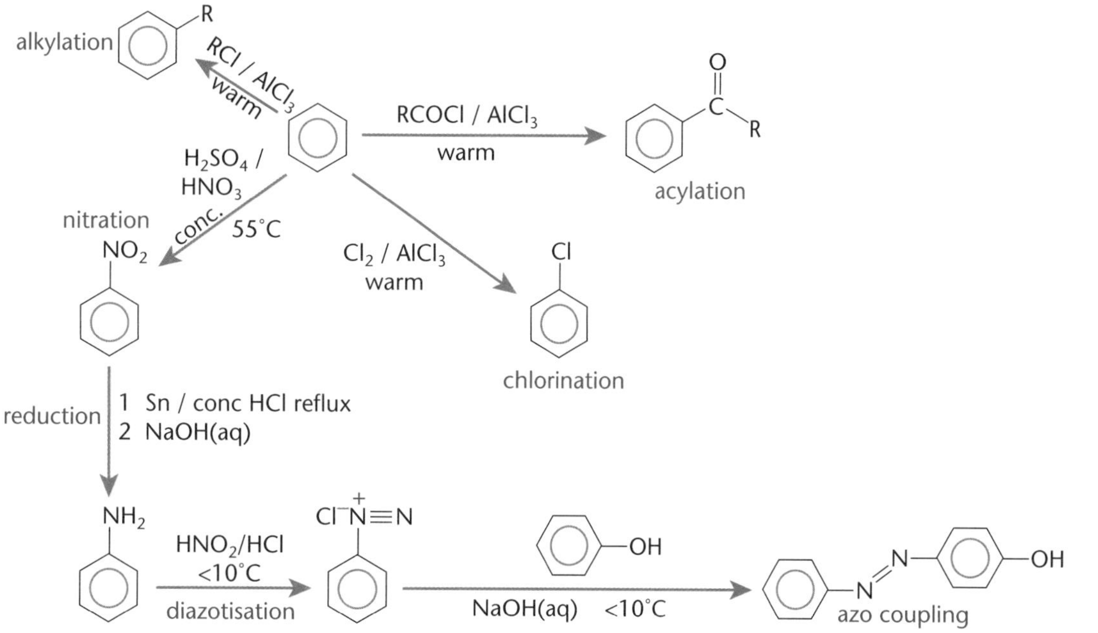

It is essential, if you are to answer such questions, that you thoroughly learn suitable reagents and conditions for all the reactions in your syllabus.

- Questions are often set in exams asking for reagents, conditions or products.
- More searching problems may expect a synthetic route from a starting material to a final product. The synthetic route may include several stages.

Synthesis of medicines

OCR M4

Synthesis of chiral drugs

The structures of many drugs include a chiral carbon atom. Often the drug is **stereo-specific** with only one of the optical isomers having the desired effect.

Modern synthesis of a pharmaceutical is often carried out to produce a **single optical isomer**. This **reduces** any **risks** on the body from **side effects** of the 'other' optical isomer.

It is difficult to separate optical isomers produced by chemical synthesis and this increases costs. Various strategies have been developed to synthesise just the required optical isomer.

In the synthesis of stereo-specific drugs, it is important to understand the mechanism of the reaction and how this can help to plan the synthesis.

- **Enzymes** or bacteria could be used for any synthetic steps that introduce a chiral centre. Enzymes promote stereo-selectivity.
- Methods of chemical synthesis have been devised that use '**chiral catalysts**'.
- Natural chiral molecules, such as L-amino acids or D-sugars, can be used as starting materials for the synthesis. This makes use of the natural '**chiral pool**'.

Combinatorial chemistry

SALTERS M5

More effective medicines can be obtained by modifying the structure of existing medicines. **Combinatorial chemistry** is a modern technique that has been adopted by the pharmaceutical industry in drug research.

In combinatorial chemistry a **large number** of related compounds are made together so that their potential effectiveness as medicines can be assessed by **large-scale screening**.

This process often includes passing reactants over reagents on polymer supports.

Testing and identifying a new medicine

Testing a medicine involves clinical trials to assess a potential new drug:

- Step I – Is it safe?
- Step II – Does it work?
- Step III – Is it better than the standard treatment?

Chemists have a role in:

- designing and making new compounds for use as pharmaceuticals
- ethical testing
- using computer modelling in the design of medicines.

Pharmacologically active compounds

Many medicines and drugs are **pharmacologically active** – they induce a biological interaction of a living organism.

The structure and action of a pharmacologically active compound depends upon:

- the **pharmacophore** and groups that modify it
- its **interaction** with receptor sites
- the ways that species interact in three dimensions (size, shape, bond formation, orientation).

Identification of the pharmacophore is a key step in devising a synthesis for a new drug.

The pharmacophore is the part of a molecular structure that is responsible for a particular biological or pharmacological interaction.

Retrosynthesis

SALTERS M5

Retrosynthesis is a modern method for devising an organic synthesis.

Conventional synthetic routes are developed by working forwards from starting materials to a target molecule.

- In retrosynthesis an **alternative synthesis** is worked out, **backwards** from a target molecule into simpler structures.
- This procedure is repeated until simple or commercially available starting materials are reached.

Retrosynthesis is useful for identifying **potential synthetic routes** in a logical and straightforward fashion.

The chemist will then **reverse** this process to develop a synthetic route towards the required target molecule.

Sample question and model answer

The structures of the four esters with the molecular formula $C_4H_8O_2$ are shown below.

A	B	C	D
$CH_3CH_2COOCH_3$	$CH_3COOCH_2CH_3$	$HCOOCH_2CH_2CH_3$	$HCOOCH(CH_3)_2$

(a) For each structure predict the number of peaks in its NMR spectrum and the ratio of the number of protons responsible for each peak.

When we are predicting the number of peaks, we are identifying the number of different types of proton.

In compound **D**, $HCOOCH(CH_3)_2$, both the methyl groups (6H) are equivalent and will have the same chemical shift.

Ester A, $CH_3CH_2COOCH_3$, has 3 peaks ✓ in the ratio 3:2:3 ✓

Ester B, $CH_3COOCH_2CH_3$, has 3 peaks ✓ in the ratio 3:2:3 ✓

Ester C, $HCOOCH_2CH_2CH_3$, has 4 peaks ✓ in the ratio 1:2:2:3 ✓

Ester D, $HCOOCH(CH_3)_2$, has 3 peaks ✓ in the ratio 1:1:6 ✓ [8]

(b) The NMR spectrum of one of the four esters is shown below. Integration data is shown by each peak. Analyse the spectrum to find out which ester has produced it. Include all the supporting evidence from the spectrum.

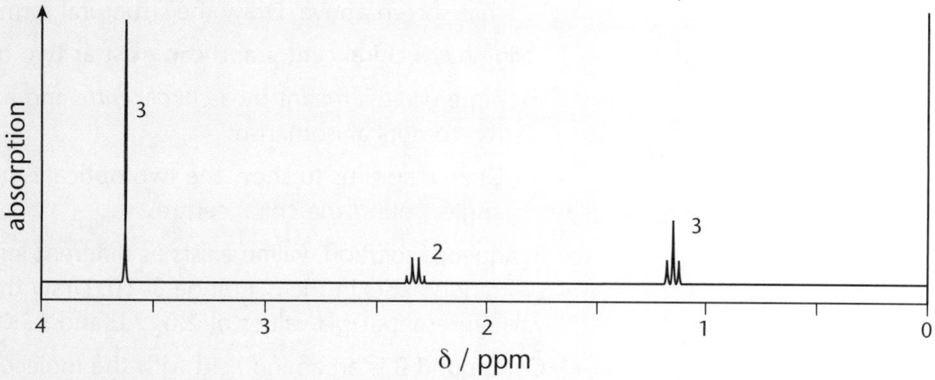

Notice how the answer is methodical, going through each piece of evidence separately.

The combination of a triplet and quartet is a giveaway for a CH_3CH_2 combination.

The integration data supports either structure A or B:

There are three types of proton in the ratio 3:2:3 ✓

The splitting pattern supports either structure A or B:

The CH_3 group at δ = 1.14 ppm has been split into a triplet ✓ by 2 protons on an adjacent carbon atom. ✓

The CH_2 group at δ = 2.29 ppm has been split into a quartet ✓ by 3 protons on an adjacent carbon atom. ✓

The integration data and splitting patterns point to either structure **A** or **B**.

The chemical shifts are conclusive though.

The CH_3 group at δ = 3.68 ppm is a singlet ✓ so there can be no protons on an adjacent carbon. ✓

From chemical shifts,

the CH_3 group at δ = 3.68 ppm must be adjacent to an O atom ✓

the CH_2 group at δ = 2.29 ppm must be adjacent to a C=O group. ✓

Taking all the evidence together:

the ester must be A, $CH_3CH_2COOCH_3$. ✓ [10]

(c) The IR and mass spectra can also be used to help confirm structures. What key IR absorptions and *m/z* values would you expect to see in the mass and IR spectra of ester **A**?

There are 4 marks here. The first is for the key information: the C=O absorption in the IR and the molecular ion peak in the mass spectrum. The second mark is for identification of another key feature.

IR: A C=O absorption at about 1700 cm^{-1}. ✓ A C–O absorption at 1000–1300 cm^{-1}. There should be no absorption above 3000 cm^{-1} as there is no –OH group present. ✓

Other fragment ions could have been chosen.

Mass spectrum: A molecular ion peak at m/z = 88 confirming the molecular mass. ✓ Fragments ions at m/z = 57 from $CH_3CH_2CO^+$; also at m/z = 31 for CH_3O^+. ✓ [4]

[Total: 22]

Practice examination questions

1 Valine, $(CH_3)_2CH(NH_2)COOH$, is an amino acid.

(a) What is the 'R' group in valine? [1]

(b) Valine reacts with an amino acid, **A**, to form the dipeptide below.

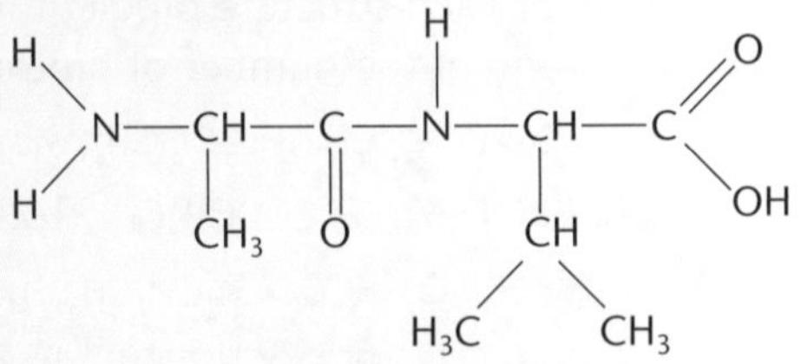

(i) Draw a circle around the peptide linkage.

(ii) Draw the structure of the amino acid, **A**.

(iii) Valine can react with the amino acid, **A**, to form a different dipeptide from that shown above. Draw the structural formula of this other dipeptide. [3]

(c) Valine has a chiral centre and can exist as two optical isomers.

(i) State what is meant by a *chiral centre* and explain how a chiral centre gives rise to optical isomerism.

(ii) Draw diagrams to show the two optical isomers of valine. State the bond angle around the chiral centre. [5]

(d) In aqueous solution, valine exists as different ions at different pH values. The zwitterion exists in the pH range 3–10. Draw the displayed formula of the ion of valine present at pH values of 2.0, 7.0 and 12.0. [3]

(e) Compound **B** is an amino acid with the molecular formula $C_3H_7NO_3$.

Draw the structure of amino acid **B**. [1]

[Total: 13]

2 Lactic acid has the structural formula $CH_3CH(OH)COOH$.

(a) What is the systematic name for lactic acid? [1]

(b) Lactic acid has optical isomers.

(i) What structural feature in lactic acid results in optical isomerism?

(ii) How does optical isomerism arise?

(iii) Draw a three-dimensional diagram to show the optical isomers of lactic acid. [4]

(c) Lactic acid can polymerise to form poly(lactic acid)

(i) What type of polymer is poly(lactic acid)?

(ii) Draw a short section of poly(lactic acid) showing two repeat units. [3]

[Total: 8]

Practice examination questions (continued)

3 The repeat units of two polymers **C** and **D** are shown below.

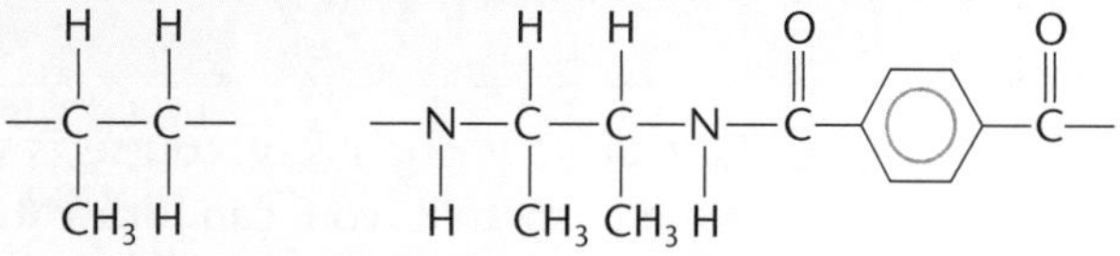

C **D**

(a) (i) Draw the structure of the monomer of polymer **C**.

(ii) Name the type of polymerisation. [2]

(b) (i) Draw the structure of the two monomers of polymer **D**.

(ii) Name the type of polymerisation. [3]

(c) Suggest why polymer **D** would be more likely to be hydrolysed than polymer **C**. [2]

[Total: 7]

4 Compounds **E** and **F** are both diols with the molecular formula $C_4H_{10}O_2$.

(a) The proton NMR spectrum of a diol **E**, $C_4H_{10}O_2$, is shown below. The numbers by each peak represent the integration trace. The peak at $\delta = 3.65$ ppm disappears when the spectrum is run a second time with D_2O added.

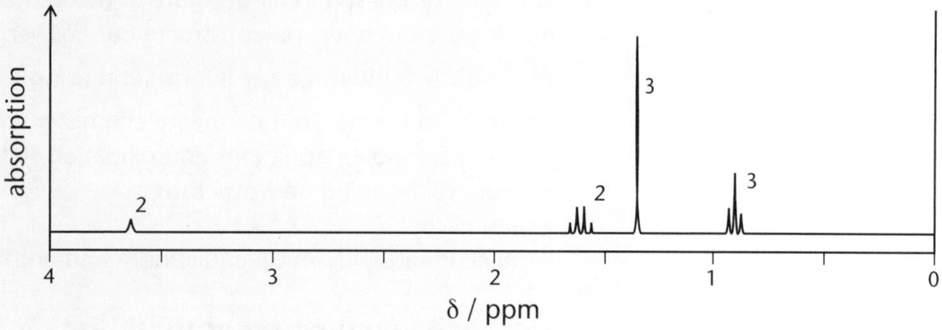

(i) How many different types of proton are present in the diol?

(ii) The peaks at $\delta = 1.63$ ppm and $\delta = 0.90$ ppm result from a single alkyl group. Identify this group and explain the splitting pattern.

(iii) What can be deduced from the single peak at $\delta = 1.32$ ppm?

(iv) What can be concluded about the peak at $\delta = 3.65$ ppm?

(v) Show the structure of the diol **E**. [9]

(b) The proton NMR spectrum diol **F**, run in D_2O, is shown below.

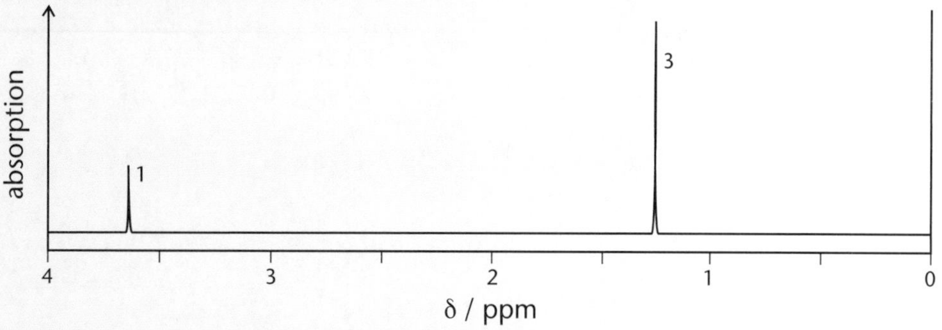

(i) What can be concluded about the two peaks?

(ii) Show the structure of the diol **F**. [5]

[Total: 14]

Chapter 6

Synoptic assessment

What is synoptic assessment?

Synoptic assessment emphasises your understanding and your application of the principles included in your chemistry course.

Part of your chemistry course is assessed using **synoptic questions**. These are written so that you can **draw together** knowledge, understanding and skills learned in **different parts** of AS and A2 Chemistry.

Synoptic assessment takes place mainly in **module exams** but part may be assessed in coursework.

What type of questions will be asked?

You make the links between different areas of chemistry yourself. You choose and use the context for your answer.

The context has been made for you. Often this will be a situation or will involve data that you will not have seen before.

You will need to answer **two** main types of synoptic question.

1 **You make links** and **use connections** between different areas of chemistry. Some examples are given below.

Using examples drawn from different parts of your chemistry course:
- discuss the role of a lone pair in chemistry…
- discuss the chemistry of water…
- compare the common types of chemical bonding…

2 **You use ideas and skills** which **permeate chemistry**.
This type of question will be more structured and you are unlikely to have as much choice in how you construct your answer.

Your task is to interpret any information using the 'big ideas' of chemistry.

Examples of themes that permeate chemistry are shown below:
- formulae, moles, equations and oxidation states
- chemical bonding and structure
- periodicity
- reaction rates, chemical equilibrium and enthalpy changes.

How will you gain synoptic skills?

A synoptic question may present you with a new piece of information.

You may be expected to calculate formulae from data, write correct formulae, balance equations and perform quantitative calculations using the mole concept, use knowledge of the Periodic Table to predict reactions of unfamiliar elements or compounds, etc.

Luckily chemistry is very much a synoptic subject. Throughout your study of chemistry, you apply many of the ideas and skills learnt during AS Chemistry or your GCSE course.

In studying A2 Chemistry, you will have been using synoptic skills naturally, probably without realising it. You certainly can make little real progress in chemistry without a sound grasp of concepts such as equations, the mole, structure and bonding!

Worked synoptic sample question

Here, some of the big ideas of chemistry are being tested: formulae, bonding, redox, moles, equations. These will always appear on an exam paper with synoptic questions.

You have nothing to fear from this type of question provided that your chemistry is sound.

Structured type question

In its reactions, sulfuric acid, H_2SO_4, can behave as an acid, an oxidising agent and as a dehydrating agent.

The displayed formula of sulfuric acid is shown below.

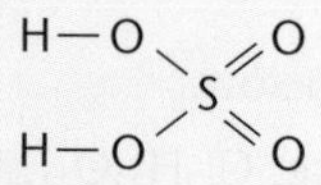

Nice clues in the question.

You are given the structure of sulfuric acid for a reason. You are being tested on whether you can apply hydrogen bonding to a new situation.

This is not much different from water's H-bonding!

(a) The boiling point of sulfuric acid is higher than expected.

Suggest and explain why the boiling point of sulfuric acid is higher than expected.

Sulfuric acid molecules form hydrogen bonds ✓

hydrogen bonds break (on boiling) ✓ [2]

Just keep your nerve when writing equations. You can choose any reasonable examples.

Do think about ionic charges. Many students blindly assume that a sulfate ion has a 1– charge but how can this be so if we have H_2SO_4 with $2H^+$ ions?

(b) Dilute sulfuric acid takes part in the typical acid reactions, reacting with carbonates, metals and alkalis.

Write balanced equations for reactions of sulfuric acid with

(i) an alkali $2NaOH + H_2SO_4 \longrightarrow Na_2SO_4 + 2H_2O$

(ii) a metal $Mg + H_2SO_4 \longrightarrow MgSO_4 + H_2$

(iii) a carbonate $MgCO_3 + H_2SO_4 \longrightarrow MgSO_4 + CO_2 + H_2O$ [3]

In (i), you are just using oxidation number rules.

– this should be 2 easy marks.

(ii) is far harder.

You have to make sure that there is the same oxidation number change up and down.

Here, each S goes down by 8 from +6 to – 2.

Each I goes up by 1 from –1 to 0 … so we need 8 x I

Finally balance the H_2O and balance the H^+.

(c) Concentrated sulfuric acid oxidises some halide ions to form the halogen.

The unbalanced equation below represents the oxidation of iodide ions by sulfuric acid.

$H^+ + SO_4^{2-} + I^- \longrightarrow I_2 + H_2S + H_2O$

(i) Determine the oxidation numbers of sulfur and iodine on either side of the equation.

SO_4^{2-} : +6 $\longrightarrow$ H_2S: -2 ✓

I^-: -1 $\longrightarrow$ I_2: 0 ✓ [2]

(ii) Balance the equation.

$10H^+ + SO_4^{2-} + 8I^- \longrightarrow 4I_2 + H_2S + 4H_2O$ ✓ [1]

Worked synoptic question (continued)

Here you are provided with information and you must use your chemical knowledge and understanding to solve the problem.

Use the information – the whole question is about **dehydration** (loss of water) caused by sulfuric acid.

Good advice is to look for the obvious. If your answer looks like some chemistry that you have never seen before, then you have probably made a mistake!

Key points here:

X is black. What do you know that is black?

Y must be made out of H, C, or O.

Z is hard but is still based on dehydration.

(d) Concentrated sulfuric acid dehydrates many organic compounds, forming water as one of the products.

For example, sulfuric acid dehydrates propan–1–ol by eliminating water to form propene.

$CH_3CH_2CH_2OH \longrightarrow CH_3CH{=}CH_2$

Three other examples are shown below.

- Sulfuric acid dehydrates sucrose, $C_{12}H_{22}O_{11}$, to form a black solid, **X**.
- Sulfuric acid dehydrates methanoic acid to form a gas, **Y**, with the same relative molecular mass as ethene.
- Sulfuric acid dehydrates ethane-1,2-diol to form a compound **Z** with a relative molecular mass of 88.0

Suggest the identity of **X**, **Y** and **Z**. Write equations for each reaction and deduce the structural formula of compound **Z**.

X: C ✓ $C_{12}H_{22}O_{11} \longrightarrow 12C + 11H_2O$ ✓

Y: CO ✓ $HCOOH \longrightarrow CO + H_2O$ ✓

Z: $C_4H_8O_2$ ✓ $2C_2H_6O_2 \longrightarrow C_4H_8O_2 + 2H_2O$ ✓

Structure:

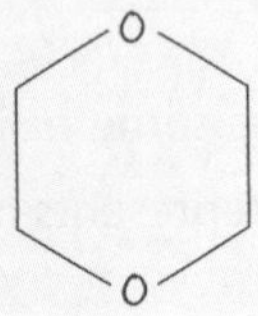

✓

[7]

[Total: 15]

Practice examination questions

1 Organic acids occur widely in nature.

(a) Butanoic acid, $CH_3(CH_2)_2COOH$, and compound **W** are straight-chain organic acids present in sweat.

(i) Compound **W** was analysed and was found to have the percentage composition by mass:

C, 66.7%; H, 11.1%; O, 22.2%. $M_r = 144.0$

Determine the molecular formula of compound **W** and suggest its structural formula.

(ii) Dogs can track humans from the odours in their sweat.

Sweat containing equal amounts of butanoic acid and compound **W** produces more butanoic acid vapour than vapour from compound **W**.

Suggest and explain a reason for this. [6]

Practice examination questions (continued)

(b) Compound **X** is a straight chain organic acid. A chemist analysed a sample of acid **X** by the procedure below.

The chemist first prepared a 100 cm^3 solution of **X** by dissolving 4.35 g of **X** in water.

In a titration, 10.00 cm^3 0.500 mol dm^{-3} NaOH were neutralised by exactly 8.50 cm^3 of solution **X**.

(i) Calculate the pH of the NaOH(aq) used in the titration.
$K_w = 1.00 \times 10^{-14}$ mol^2 dm^{-6}.

(ii) Use the results to calculate the molar mass of acid **X** and suggest its identity. [8]

[Total: 14]

2 *Refer to data on Page 144 and 151 for this question.*

Compounds **A** to **G** are isomers of $C_6H_{12}O_2$.

(a) Isomer **A**, $C_6H_{12}O_2$, is a neutral compound. Acid hydrolysis of **A** forms compounds **X** and **Y**.

X and **Y** can also both be formed from propanal by different redox reactions.

X has an absorption in its IR spectrum at 1750 cm^{-1}.

Deduce the structural formulae of **A**, **X** and **Y**. Give suitable reagents, in each case, for the formation of **X** and **Y** from propanal and state the role of the acid in the hydrolysis of **A**. [8]

(b) Isomers **B**, **C**, **D** and **E** are acidic.

B, **C** and **D** are structural isomers that also have optical isomers.

In its proton NMR spectrum, **E** has three singlets. Deduce the structural formulae **B**, **C**, **D** and **E**. [4]

(c) Isomer **F**, $C_6H_{12}O_2$, has the structural formula shown below, on which some of the protons have been labelled.

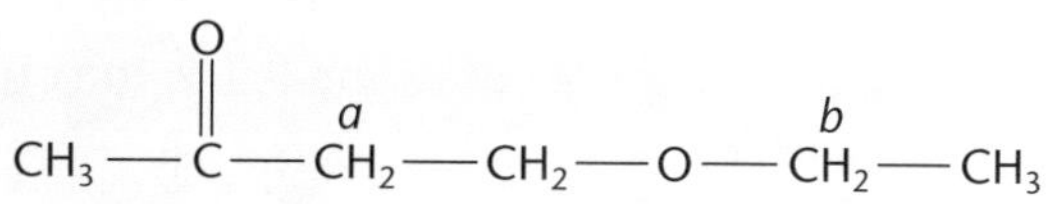

A proton NMR spectrum is obtained for **F**. Predict the chemical shift for the protons labelled *a* and *b*. Explain the splitting patterns arising from protons *a* and *b*. [6]

(d) Isomer **G**, $C_6H_{12}O_2$, contains six carbon atoms in a ring. It has an IR absorption at 3270 cm^{-1} and shows only three peaks in its proton NMR spectrum. Deduce a structural formula for **G**. [2]

[Total: 20]

Practice examination answers

Chapter 1 Rates and equilibria

1 (a) (i) 1st order with respect to H_2O_2 ✓
Double concentration of H_2O_2, rate doubles ✓
1st order with respect to I^- ✓
7 times concentration of I^-, rate x 7 ✓
Zero order with respect to H^+ ✓
Double concentration of H^+, rate stays constant ✓

(ii) Rate = $k[H_2O_2][I^-]$ ✓
2.8×10^{-2} $dm^3\ mol^{-1}\ s^{-1}$ ✓✓ [9]

(b) (i) The slowest step of a multi-step process ✓

(ii) $H_2O_2 + I^- \longrightarrow H_2O + IO^-$ ✓ [2]

[Total: 11]

2 (a) (i) Rate = $k[A]^2[B]$ ✓

(ii) 3 ✓

(iii) 27 ✓

(iv) $k = 3.14 \times 10^{-3}$ $dm^6\ mol^{-2}\ s^{-1}$ ✓✓ [5]

(b) (i) H^+ is a catalyst ✓
H^+ appears in the rate equation but not the overall equation ✓

(ii) $CH_3COCH_3(aq) + H^+(aq) \longrightarrow [CH_3COHCH_3(aq)]^+$ ✓ [3]

[Total: 8]

3 (a) $$K_c = \frac{[CH_3COOH]\ [CH_3CH_2OH]}{[CH_3COOCH_2CH_3]\ [H_2O]}$$ ✓

$K_c = 0.26$ ✓; No units ✓ [3]

(b) (i) Equilibrium would move to the left ✓ to counteract the added CH_3CH_2OH ✓

(ii) No effect ✓ K_c only changes with temperature ✓ [4]

[Total: 7]

4 (a) (i) $[H_2(g)]$, 0.44 mol dm^{-3}; ✓ $[I_2(g)]$, 0.02 mol dm^{-3}; ✓
$[HI(g)]$, 0.32 mol dm^{-3} ✓

(ii) $K_c = \frac{[HI(g)]^2}{[H_2(g)]\ [I_2(g)]}$ ✓ $= \frac{0.32^2}{0.44 \times 0.02} = 11.6$ ✓ no units ✓ [6]

b) Equilibrium moves to left ✓; Forward reaction is exothermic (or the reverse reaction is endothermic) ✓ [2]

[Total: 8]

5 (a) Proton donor ✓ [1]

(b) $HCOOH + HNO_3 \rightleftharpoons HCOOH_2^+ + NO_3^-(aq)$ ✓
HCOOH is a base because it accepts a proton ✓ [2]

(c) (i) pH = 0.75 ✓

(ii) $K_w = [H^+(aq)][OH^-(aq)]$ ✓
$[H^+(aq)] = 1.0 \times 10^{-14}/0.372 = 2.69 \times 10^{-14}$ mol dm^{-3} ✓ pH = 13.57 ✓
(iii) $K_a = \frac{[H^+(aq)]\,[HCOO^-(aq)]}{[HCOOH(aq)]} = \frac{[H^+(aq)]^2}{[HCOOH(aq)]}$ ✓
$[H^+(aq)] = \sqrt{(1.6 \times 10^{-4} \times 0.263)} = 6.49 \times 10^{-3}$ mol dm^{-3} ✓ pH = 2.19 ✓
(iv) $[H^+(aq)] = K_a \times [HCOOH(aq)]/[HCOO^-(aq)]$ ✓
$[H^+(aq)] = 1.6 \times 10^{-4} \times 0.255/0.325 = 1.26 \times 10^{-4}$ mol dm^{-3} ✓
pH = 3.90 ✓ [10]

[Total: 13]

6 (a) (i) A strong acid completely dissociates to donate protons.
A weak acid partially dissociates to donate protons. ✓
(ii) Proton acceptor ✓ [2]
(b) (i) $[OH^-(aq)] = 1.50 \times 10^{-3} \times 1000/25 = 0.060$ mol dm^{-3} ✓
$K_w = [H^+(aq)][OH^-(aq)]$ ✓
$[H^+(aq)] = 1.0 \times 10^{-14}/0.060 = 1.67 \times 10^{-13}$ mol dm^{-3} ✓ pH = 12.78 ✓
(ii) Amount of HCl added = 0.0125 mol ✓
Amount of HCl remaining = $0.0125 - 1.50 \times 10^{-3} = 0.011$ mol ✓
[HCl] = $0.011 \times 1000/75 = 0.147$ mol dm^{-3} ✓ pH = 0.83 ✓ [8]
(c) $[H^+(aq)] = 10^{-pH} = 0.0302$ mol dm^{-3} ✓
amount of HCl = $0.0302 \times 25/1000 = 7.55 \times 10^{-4}$ mol ✓
$[H^+(aq)]$ in diluted solution = $7.55 \times 10^{-4} \times 1000/40 = 0.0189$ mol dm^{-3} ✓
pH = 1.72 ✓ [4]

[Total: 14]

Chapter 2 Energy changes in chemistry

1 (a) (i)

definition	*letter*
1st ionisation energy of potassium	B
1st electron affinity of bromine	F ✓
the enthalpy of atomisation of potassium	C
the enthalpy of atomisation of bromine	A ✓
the lattice enthalpy of potassium bromide	E
the enthalpy of formation of potassium bromide	D ✓

(ii) –394 – (89 + 419 + 112 –325) ✓ = –689 kJ mol^{-1} ✓ [5]
(b) (i) KI is less exothermic because I^- has a smaller charge density ✓ and attraction between K^+ and I^- ions is weaker ✓
(ii) $CaBr_2$ is more exothermic because Ca^{2+} has a greater charge density ✓ and attraction between Ca^{2+} and Br^- ions is stronger ✓ [4]

[Total: 9]

2 (a) (i) $Mg^{2+}(g) + 2Cl^-(g) \longrightarrow MgCl_2(s)$ ✓
(ii) $Mg^{2+}(g) + aq \longrightarrow Mg^{2+}(aq)$ ✓
(iii) $MgCl_2(s) + aq \longrightarrow Mg^{2+}(aq) + 2Cl^-(aq)$ ✓ [3]
(b) ΔH(solution) = –(lattice enthalpy) + Σ(enthalpy changes of hydration) ✓
= –(–2526) + (–1891 + 2 × –384) ✓ = –133 kJ mol^{-1} ✓ [3]

[Total: 6]

3 $\Delta H_r = \Sigma\Delta H_f(\text{products}) - \Sigma\Delta H_f(\text{reactants})$
$= (-1676) - (-602)$ ✓
$= -1074\ \text{kJ mol}^{-1}$ ✓
$\Delta S_r = \Sigma S\ (\text{products}) - \Sigma S\ (\text{reactants})$
$= [(2 \times 81) + 51] - [27 + (2 \times 28)]$ ✓
$= 130\ \text{J K mol}^{-1}$ ✓ $= 0.13\ \text{kJ K}^{-1}\ \text{mol}^{-1}$ ✓
$\Delta G = \Delta H - T\Delta S$ ✓
$\Delta G = -1074 - 298 \times 0.13 = -1113\ \text{kJ mol}^{-1}$ ✓
Reaction is feasible at 298 K because $\Delta G < 0$ ✓

[Total: 8]

4 (a) 298 K; $[Ni^{2+}(aq)]$ 1 mol dm^{-3} ✓; $[Fe^{2+}(aq)]$ and $[Fe^{3+}(aq)]$ are both 1 mol dm^{-3} ✓ [2]
(b) It allows ions to flow between half-cells ✓ [1]
(c) (i) 1.02 V ✓
(ii) Electrons flow along wire from nickel half-cell ✓
(iii) Ni is – electrode; Pt is + electrode ✓ [3]
(d) (i) $Ni \longrightarrow Ni^{2+} + 2e^-$ ✓
$Fe^{3+} + e^- \longrightarrow Fe^{2+}$ ✓
(ii) $2Fe^{3+} + Ni \longrightarrow 2Fe^{2+} + Ni^{2+}$ ✓ [3]
(e) In Ni redox equilibrium, increase in $[Ni^{2+}(aq)]$ shifts equilibrium to the right.
Electrons are less available and electrode potential becomes less negative. ✓
Difference in cell potentials becomes less and cell potential will decrease. ✓ [2]

[Total: 11]

5 (a) (i) $Mn^{3+}(aq)$ ✓
(ii) $Mn^{3+}(aq)$ ✓ I^- must provide electrons and have the more negative $E^\ominus$ (+0.54). ✓ The more positive $E^\ominus$ system (+1.49 V) will gain electrons. ✓ [4]
(b) 0.66 V ✓
$4V^{2+}(aq) + O_2(g) + 2H_2O(l) \longrightarrow 4OH^-(aq) + 4V^{3+}(aq)$ ✓✓ [3]

[Total: 7]

Chapter 3 The Periodic Table

1 (a) Ni: $1s^2 2s^2 2s^6 3s^2 3p^6 3d^8 4s^2$ ✓; Ni^{2+}: $1s^2 2s^2 2s^6 3s^2 3p^6 3d^8$ ✓ [2]
(b) A d-block element has its highest energy electron in a d sub-shell. ✓
A transition element has at least one ion with a partially filled d sub-shell. ✓ [2]
(c) (i) Lone pair of electrons on oxygen atom ✓
(ii) Ion $[Ni(H_2O)_6]^{2+}$ ✓
(iii) Covalent bonding in H_2O ✓ and coordinate bonding from water ligands to the Ni^{2+} ion. ✓
(iv) Octahedral ✓ ; 90° ✓ [6]
(d) (i) $Ni(OH)_2$ or $Ni(OH)_2(H_2O)_4$ ✓
$Ni^{2+}(aq) + 2OH^-(aq) \longrightarrow Ni(OH)_2(s)$ ✓
or $[Ni(H_2O)_6]^{2+}(aq) + 2OH^-(aq) \longrightarrow Ni(OH)_2(H_2O)_4(s) + 2H_2O(l)$
(ii) $[NiCl_4]^{2-}$ ✓
(iii) $[Ni(H_2O)_6]^{2+}(aq) + 6NH_3(aq) \longrightarrow [Ni(NH_3)_6]^{2+}(aq) + 6H_2O(l)$ ✓
ligand substitution ✓ [5]

[Total: 15]

2 (a) (i) +6 ✓

(ii) +1 ✓

(iii) +4 ✓ [3]

(b) $2Cr^{3+} + 10OH^- + 3H_2O_2 \longrightarrow 2CrO_4^{2-} + 8H_2O$ ✓ [1]

(c) Vanadium as V_2O_5 in the contact process for the manufacture SO_3 for H_2SO_4 ✓
Iron in the Haber process for the manufacture of NH_3 ✓ [2]

[Total: 6]

3 (a) (i) +7 ✓

(ii) $5H_2O_2 + 2MnO_4^- + 6H^+ \longrightarrow 5O_2 + 8H_2O + 2Mn^{2+}$ ✓ [2]

(b) Reducing agent ✓ [1]

(c) Amount of $KMnO_4 = 0.0150 \times \frac{23.80}{1000} = 3.57 \times 10^{-4}$ mol ✓

Amount of H_2O_2 that reacted $= 2.5 \times 3.57 \times 10^{-4} = 8.925 \times 10^{-4}$ mol ✓

Concentration of $H_2O_2 = \frac{1000}{25.0} \times 8.925 \times 10^{-4} = 0.0357$ mol ✓ [3]

[Total: 6]

4 (a) (i) A coordinate (dative covalent bond) forms ✓ between a lone pair on a ligand and the central metal ion. ✓

(ii) Bidentate ligand ✓ [3]

(b) (i) +3 ✓

(ii) $1s^22s^22s^63s^23p^63d^6$ ✓ [2]

(c) (i) 6 ✓

(ii) Octahedral ✓

(iii) Add an excess of 1,2-diaminoethane or $NH_2CH_2CH_2NH_2$ ✓
$[Ni(H_2O)_6]^{2+} + 3NH_2CH_2CH_2NH_2 \rightarrow [Ni(NH_2CH_2CH_2NH_2)_3]^{2+} + 6H_2O$ ✓ [4]

[Total: 9]

Chapter 4 Chemistry of organic functional groups

1 (a) C : H : O = 66.7/12 : 11.1/1 : 22.2/16 ✓
Molecular formula = C_4H_8O ✓ [2]

(b) $CH_3CH_2COCH_3$ ✓; $CH_3CH_2CH_2CHO$ ✓; $(CH_3)_2CHCHO$ ✓ [3]

(c) 2,4-dinitrophenylhydrazine ✓ orange precipitate ✓ [2]

(d) Tollens' reagent ✓. Reacts with butanal and methylpropanal ✓
Silver mirror forms ✓ (or $H_2SO_4/K_2Cr_2O_7$; from orange to green) [3]

(e) Product: $CH_3CH_2CH(OH)CH_3$ ✓ formed from $CH_3CH_2COCH_3$ (or Fehling's or Benedict's solution gives a red precipitate) ✓ [2]

[Total: 12]

2 (a) (i) Ethyl propanoate ✓

(ii) **E**: CH_3CH_2OH ✓
F: $CH_3CH_2COO^-Na^+$ ✓✓ (1 mark if CH_3CH_2COOH) [4]

(b) (i)

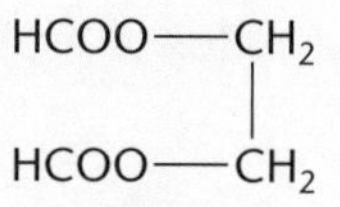

✓

(ii) Concentrated H_2SO_4 ✓ [2]

(c)

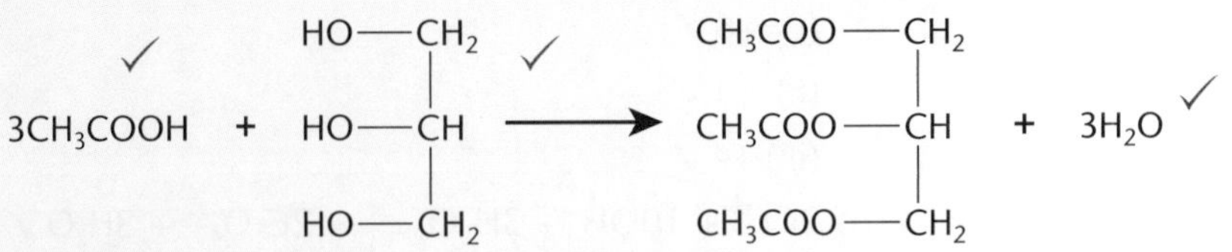

[3]

[Total: 9]

3 (a) (i) Molar ratio C:H = 90.56/12 : 9.44/1 ✓

Empirical formula = C_4H_5 ✓

Molecular formula = C_8H_{10} ✓

(ii)

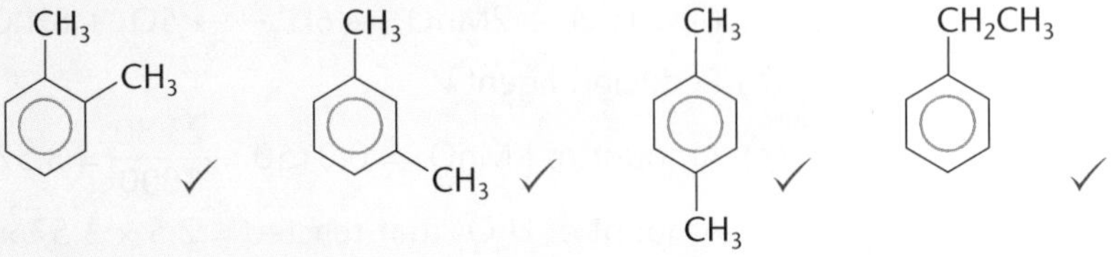

[7]

(b) (i) 1,4-dimethylbenzene ✓

(ii)

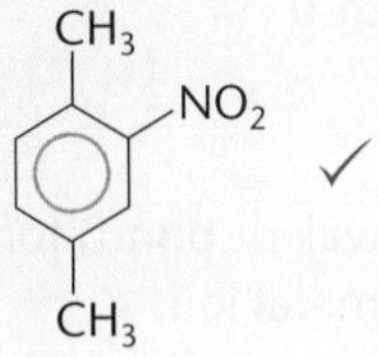

[2]

[Total: 9]

4 (a) (i) Electrophilic substitution ✓

(ii) Concentrated HNO_3 and concentrated H_2SO_4 ✓ [2]

(b) (i) Tin and concentrated HCl ✓

(ii)

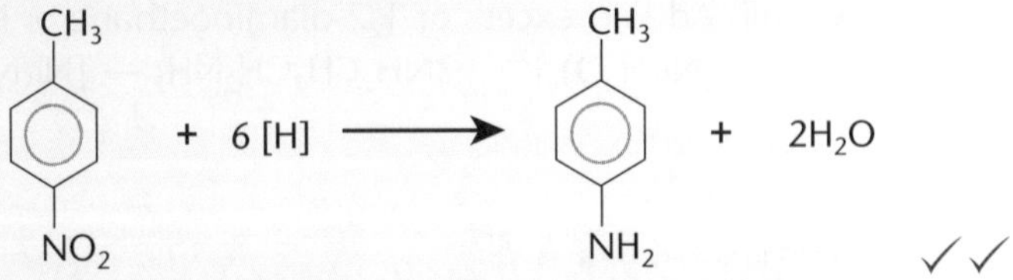

[3]

(c) (i) Reagents $NaNO_2$/HCl(aq) ✓ Conditions: <10°C ✓

(ii)

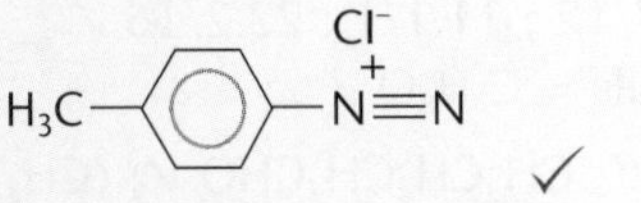

[3]

(d) (i) A phenol, e.g. C_6H_5OH ✓

(ii)

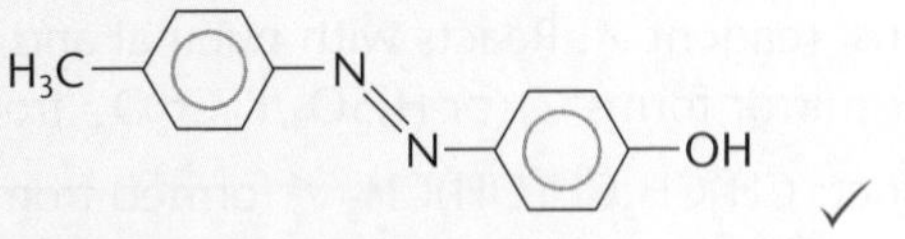

[2]

[Total: 10]

5 (a) Compound **A**: $CH_3CH_2CH_2COOH$ ✓

Compound **B**: $(CH_3)_2CHOH$ ✓ [2]

(b) Step 1: $H_2SO_4/K_2Cr_2O_7$ ✓ reflux ✓ oxidation ✓

Step 2: $NaBH_4$ ✓ with H_2O or ethanol ✓ reduction ✓

Step 3: Conc. H_2SO_4 catalyst ✓ reflux ✓ esterification ✓ [9]

[Total: 11]

Chapter 5 Polymers, analysis and synthesis

1 (a) $(CH_3)_2CH$ ✓ [1]

(b) (i)

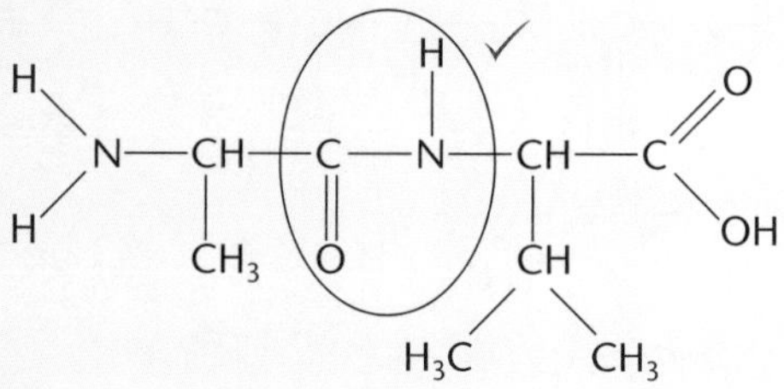

(ii)

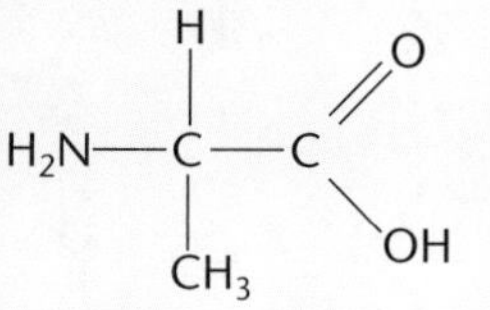 ✓

(iii)

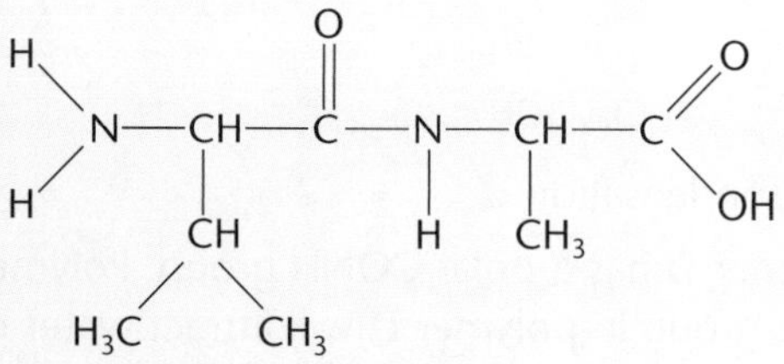 ✓ [3]

(c) (i) A chiral carbon atom has **four** different groups attached to it. ✓ It provides two possible forms that are non-superimposable mirror images of one another. ✓

(ii)

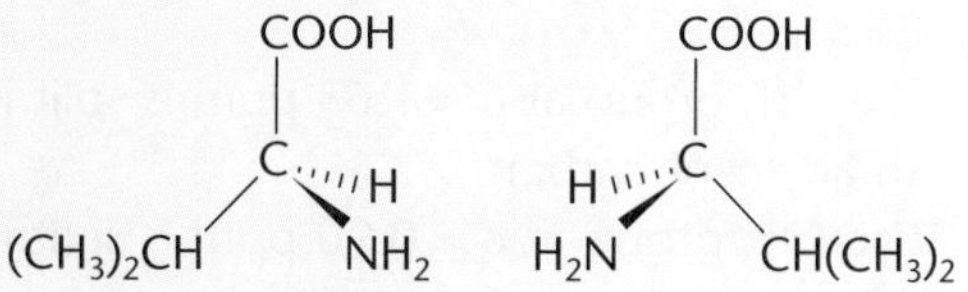 ✓✓

Bond angle = 109.5° ✓ [5]

(d)

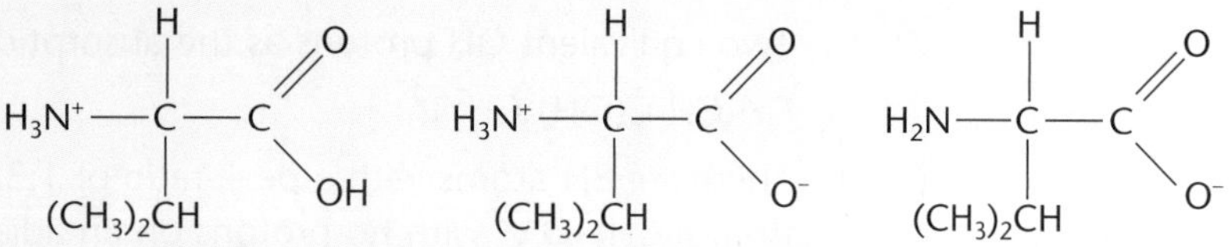

pH = 2 pH = 7 pH = 12 ✓✓✓ [3]

(e)

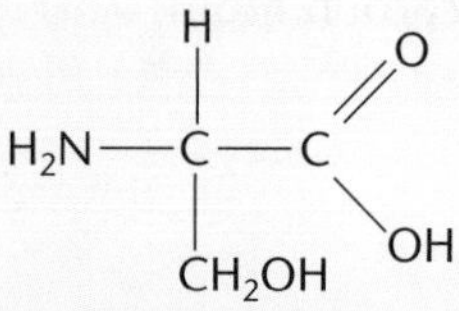 ✓ [1]

[Total: 13]

2 (a) 2-hydroxypropanoic acid ✓ [1]

(b) (i) A chiral carbon with four different groups attached ✓

(ii) Optical isomers are non-superimposable mirror images ✓

(iii)

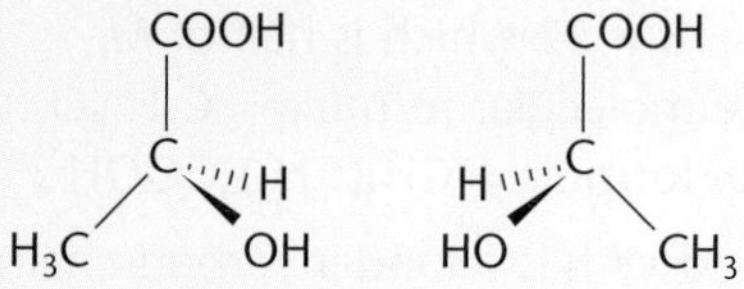 ✓✓ [4]

(c) (i) Polyester or a condensation polymer ✓

(ii) 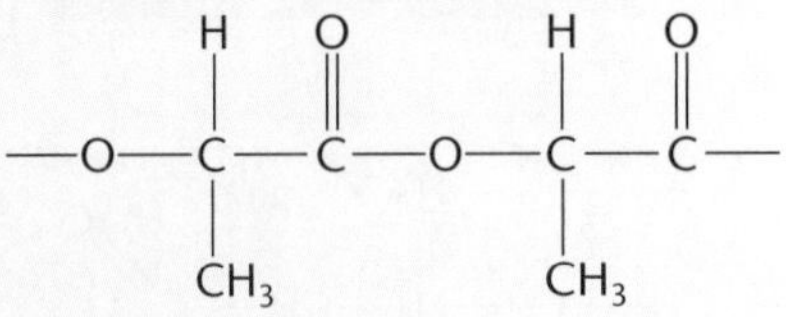

ester link ✓ structure ✓ [3]

[Total: 8]

3 (a) (i) 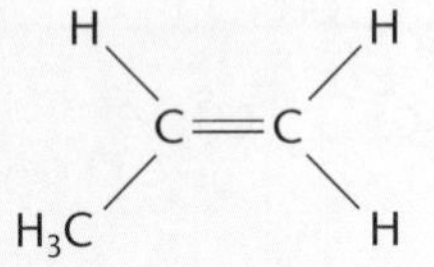 ✓

(ii) Addition ✓ [2]

(b) (i) 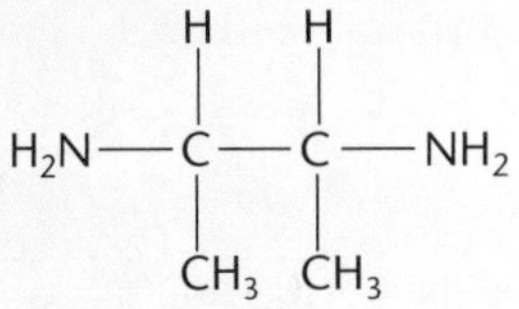 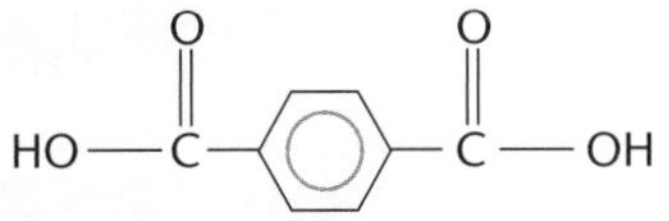

✓✓

(ii) Condensation ✓ [3]

(c) Polymer **D** has a polar CONH group. Polymer **C** is non-polar. ✓
Polar group in polymer **D** will attract water or hydroxide ions. ✓ [2]

[Total: 7]

4 (a) (i) 4 ✓

(ii) C_2H_5 ✓
The CH_2 group at δ = 1.63 ppm is split into a quartet ✓ by 3 protons on an adjacent carbon. ✓
The CH_3 group at δ = 0.90 ppm is split into a triplet ✓ by 2 protons on an adjacent carbon. ✓

(iii) A CH_3 group with no protons on an adjacent carbon. ✓

(iv) Two equivalent OH protons as the absorption disappears with D_2O. ✓

(v) $CH_3CH_2C(OH)_2CH_3$ ✓ [9]

(b) (i) There are 8H atoms with a peak ratio of 1:3. The peak at δ = 3.64 ppm is from a CH_2–O ✓ with no protons on an adjacent carbon atom ✓
The peak at δ = 1.24 ppm is from two equivalent CH_3 groups ✓ with no protons on an adjacent carbon atom. ✓

(ii) $(CH_3)_2C(OH)CH_2OH$ ✓ [5]

[Total: 14]

Chapter 6 Synoptic assessment

1 (a) (i) $C : H : O = \frac{66.7}{12} : \frac{11.1}{1} : \frac{22.2}{16}$ ✓

= 5.56 : 11.1 : 1.39 = 4 : 8 : 1

empirical formula = C_4H_8O ✓

48 + 8 + 16 = 72 which is half of M_r

Therefore molecular formula = $C_8H_{16}O_2$ ✓

Structural formula = $CH_3(CH_2)_6COOH$ ✓

(ii) compound **W** has a longer carbon chain and contains more electrons ✓
compound **W** has more van der Waals' forces (or more instantaneous dipole-induced dipole interactions). ✓ [6]

(b) (i) $[H^+(aq)] = \frac{K_w}{[OH^-(aq)]}$ ✓ $= \frac{1.00 \times 10^{-14}}{0.500}$

$= 2.00 \times 10^{-14}$ mol dm^{-3} ✓

pH = –log[H$^+$(aq)] = –log(2.00 x 10^{-14}) = 13.70 ✓

(ii) moles NaOH in 10.00 cm^3 = moles NaOH = 0.005 00 mol ✓

moles **A** in 8.50 cm^3 = moles NaOH = 0.005 00 mol ✓

moles **A** in 100 cm^3 = $\frac{0.00500 \times 100}{8.50}$ = 0.0588 mol ✓

molar mass of **A** = $\frac{4.35}{0.0588}$ = 74.0 g mol^{-1} ✓

Therefore **A** is propanoic acid / CH_3CH_2COOH ✓ [8]

[Total: 14]

2 (a) **A** = $CH_3CH_2COOCH_2CH_2CH_3$ ✓

X = CH_3CH_2COOH ✓

Y = $CH_3CH_2CH_2OH$ ✓

For propanal ⟶ **X**, H_2SO_4 ✓ and $K_2Cr_2O_7$ ✓

For propanal ⟶ **Y**, $NaBH_4$ ✓ in water ✓

Acid is a catalyst ✓ [8]

(b) **B**, **C** and **D**

$CH_3CH_2CH_2CH(CH_3)COOH$ ✓

$CH_3CH_2CH(CH_3)CH_2COOH$ ✓

$(CH_3)_2CHCH(CH_3)COOH$ ✓

E: $(CH_3)_3CCH_2COOH$ ✓ [4]

(c) *a*: δ = 3.3–4.3 ppm ✓; *y*: δ = 2.0–2.9 ppm ✓

a: splitting will be a quartet ✓ as there is an adjacent CH_3 ✓

b: splitting will be a triplet ✓ as there is an adjacent CH_2 ✓ [6]

(d)

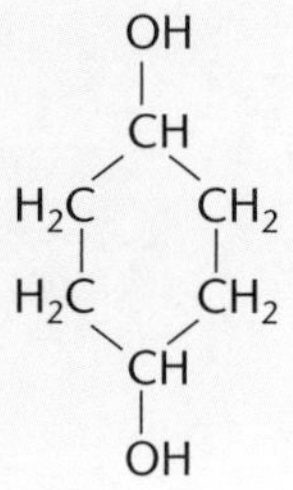

✓ (structure) ✓ (2 x OH groups at correct positions) [2]

[Total: 20]

Notes

Notes

Periodic Table

The Periodic Table

Key: relative atomic mass — 1.0; atomic symbol — H; atomic number — 1; name — Hydrogen

1	2											3	4	5	6	7	0
																	4.0 **He** 2 Helium
6.9 **Li** 3 Lithium	9.0 **Be** 4 Beryllium											10.8 **B** 5 Boron	12.0 **C** 6 Carbon	14.0 **N** 7 Nitrogen	16.0 **O** 8 Oxygen	19.0 **F** 9 Fluorine	20.2 **Ne** 10 Neon
23.0 **Na** 11 Sodium	24.3 **Mg** 12 Magnesium											27.0 **Al** 13 Aluminium	28.1 **Si** 14 Silicon	31.0 **P** 15 Phosphorus	32.1 **S** 16 Sulfur	35.5 **Cl** 17 Chlorine	39.9 **Ar** 18 Argon
39.1 **K** 19 Potassium	40.1 **Ca** 20 Calcium	45.0 **Sc** 21 Scandium	47.9 **Ti** 22 Titanium	50.9 **V** 23 Vanadium	52.0 **Cr** 24 Chromium	54.9 **Mn** 25 Manganese	55.8 **Fe** 26 Iron	58.9 **Co** 27 Cobalt	58.7 **Ni** 28 Nickel	63.5 **Cu** 29 Copper	65.4 **Zn** 30 Zinc	69.7 **Ga** 31 Gallium	72.6 **Ge** 32 Germanium	74.9 **As** 33 Arsenic	79.0 **Se** 34 Selenium	79.9 **Br** 35 Bromine	83.8 **Kr** 36 Krypton
85.5 **Rb** 37 Rubidium	87.6 **Sr** 38 Strontium	88.9 **Y** 39 Yttrium	91.2 **Zr** 40 Zirconium	92.9 **Nb** 41 Niobium	95.9 **Mo** 42 Molybdenum	[98] **Tc** 43 Technetium	101.1 **Ru** 44 Ruthenium	102.9 **Rh** 45 Rhodium	106.4 **Pd** 46 Palladium	107.9 **Ag** 47 Silver	112.4 **Cd** 48 Cadmium	114.8 **In** 49 Indium	118.7 **Sn** 50 Tin	121.8 **Sb** 51 Antimony	127.6 **Te** 52 Tellurium	126.9 **I** 53 Iodine	131.3 **Xe** 54 Xenon
132.9 **Cs** 55 Caesium	137.3 **Ba** 56 Barium	138.9 **La*** 57 Lanthanum	178.5 **Hf** 72 Hafnium	180.9 **Ta** 73 Tantalum	183.8 **W** 74 Tungsten	186.2 **Re** 75 Rhenium	190.2 **Os** 76 Osmium	192.2 **Ir** 77 Iridium	195.1 **Pt** 78 Platinum	197.0 **Au** 79 Gold	200.6 **Hg** 80 Mercury	204.4 **Tl** 81 Thallium	207.2 **Pb** 82 Lead	209.0 **Bi** 83 Bismuth	[209] **Po** 84 Polonium	[210] **At** 85 Astatine	[222] **Rn** 86 Radon
[223] **Fr** 87 Francium	[226] **Ra** 88 Radium	[227] **Ac*** 89 Actinium	[261] **Rf** 104 Rutherfordium	[262] **Db** 105 Dubnium	[266] **Sg** 106 Seaborgium	[264] **Bh** 107 Bohrium	[277] **Hs** 108 Hassium	[268] **Mt** 109 Meitnerium	[271] **Ds** 110 Darmstadtium	[272] **Rg** 111 Roentgenium							

Elements with atomic numbers 112–116 have been reported but not fully authenticated

lanthanides	140.1 **Ce** 58 Cerium	140.9 **Pr** 59 Praseodymium	144.2 **Nd** 60 Neodymium	144.9 **Pm** 61 Prcmethium	150.4 **Sm** 62 Samarium	152.0 **Eu** 63 Europium	157.2 **Gd** 64 Gadolinium	158.9 **Tb** 65 Terbium	162.5 **Dy** 66 Dysprosium	164.9 **Ho** 67 Holmium	167.3 **Er** 68 Erbium	168.9 **Tm** 69 Thulium	173.0 **Yb** 70 Ytterbium	175.0 **Lu** 71 Lutetium
actinides	232.0 **Th** 90 Thorium	[231] **Pa** 91 Protactinium	238.1 **U** 92 Uranium	[237] **Np** 93 Neptunium	[242] **Pu** 94 Plutonium	[243] **Am** 95 Americium	[247] **Cm** 96 Curium	[245] **Bk** 97 Berkelium	[251] **Cf** 98 Californium	[254] **Es** 99 Einsteinium	[253] **Fm** 100 Fermium	[256] **Md** 101 Mendelevium	[254] **No** 102 Nobelium	[257] **Lr** 103 Lawrencium

Index